Computer Simulation, Rhetoric, and the Scientific Imagination

Computer Simulation, Rhetoric, and the Scientific Imagination

How Virtual Evidence Shapes Science in the Making and in the News

Aimee Kendall Roundtree

LEXINGTON BOOKS
Lanham • Boulder • New York • Toronto • Plymouth, UK

Published by Lexington Books
A wholly owned subsidiary of Rowman & Littlefield
4501 Forbes Boulevard, Suite 200, Lanham, Maryland 20706
www.rowman.com

10 Thornbury Road, Plymouth PL6 7PP, United Kingdom

British Library Cataloguing in Publication Information Available

Library of Congress Cataloging-in-Publication Data

Roundtree, Aimee Kendall, 1972–
Computer simulation, rhetoric, and the scientific imagination : how virtual evidence shapes science in the making and in the news / Aimee Kendall Roundtree.
pages cm
Includes bibliographical references and index.
ISBN 978-0-7391-7556-9 (cloth : alk. paper)—ISBN 978-0-7391-7557-6 (electronic)
1. Computer simulation—Philosophy. 2. Science—Public opinion. 3. Rhetoric. 4. Information technology—Public opinion. 5. Virtual reality—Social aspects. I. Title.
QA76.9.C65R68 2014
006.8—dc23
2013037816

∞™ The paper used in this publication meets the minimum requirements of American National Standard for Information Sciences Permanence of Paper for Printed Library Materials, ANSI/NISO Z39.48-1992.

Printed in the United States of America

To Carlton.
To Kirby and Zachary.
To Fred, Lizz, and Scott.
To Levorgia, Oteal, and Irma.
To John and Nicole.
And to God from whom all blessings flow.

Contents

List of Figures

List of Tables

Acknowledgments

I must thank John Slatin, my mentor, for his painstaking care, patient guidance, and wise direction. I also must thank Jerome Bump, Lester Faigley, Robert Koons, Clay Spinuzzi, and Margaret Syverson for their input and careful review of prior versions of this work. This project could not have existed without the generous cooperation of Brent Iverson, Navin Varadarajan, Anthony Mezzacappa, and John Blondin, all of whom volunteered primary source materials and information. Likewise, I could not have persevered without support from my family—Kirby, Zachary, Fred and Lizz, Carlton and Lowell—for their prayers and encouragement. Thanks also to my friends—especially, Heather Hajdu Plank, Kasandra Diomande, Jennifer Wu, and Jason Crain—for their support. I owe special thanks to my friends and colleagues Jason Craft, Lynda Walsh, Doug Norman, and Nicole LaRose for reviewing earlier drafts of this work. I am grateful to Davida Charney for introducing me to the rhetoric of science. Finally, I truly appreciate the support extended to me by the University of Houston-Downtown vis-à-vis a Faculty Leave Award from the Office of the Provost and encouragement from faculty colleagues.

Chapter One

Why Computer Simulations Need Rhetorical Intervention

Yale University philosopher Nick Bostrom thinks that it is highly likely that we are currently living in a computer simulation. According to his paper published in *Philosophical Quarterly*, logic dictates that we are living in one. He uses philosophical assumptions such as substrate-independence and the indifference principle to reach the following conclusion: If it's true that we're heading toward a post-human age run by computer simulations, then there is no reason why we are not living in such a simulation right now. "Unless we are now living in a simulation, our descendants will almost certainly never run an ancestor-simulation" (Bostrom 11).

Bostrom's thesis caused quite a stir in the media and the discipline. It made news in the lifestyle sections of the *New Scientist*, the *Sunday Telegraph*, the *Village Voice*, and the *Times Higher Education Supplement*, to name a few. It also fueled a heated debate and discussion with fellow philosophers. George Mason University's Robin Hanson proposed that we could live as hedonists if everything is simulated. Cambridge University's John Barrow argued that the simulation in which we likely live explains why we might see drifts in laws of nature over time. Others say it is more likely that nature will wipe humankind out before any generation can actually produce such believable simulations. Therefore, we are living in an original age, not a copy (Brooks).

Whether or not we can ever know for sure, the curiosity bespeaks theorists' fascination with computer simulations' alleged capacity to dupe unsuspecting bystanders into believing that a replica is the real thing. In particular, rhetorical and critical theories often use simulation as a trope to help explain how postmodern communication works. However, computer simulations don't just exist in theory. They exist in truth for our examination. Contemporary scientists often use computer simulations when their research question is too complex (e.g., has too many factors, requires too much iteration, or involves too many calculations) to play out with any degree of accuracy otherwise. Current theories gloss over actual applications of computer simulation.

However, for many reasons, it is necessary to understand actual computer simulations as rhetorical acts and products. First, the field of computational science could benefit from practical analysis. Computational science is a relatively new but very pervasive scientific discipline ripe for a rhetorical intervention. Understanding the rhetorical components of computer simulations can actually help the discipline answer its two most important, lingering questions: what is computational knowledge and how do we achieve it?

Furthermore, rhetoric can lay the groundwork for more careful criticism of simulations at play in mundane and major cultural and political events. Experts count on simulations for everything from daily weather forecasts to complex investigations surrounding Hurricane Katrina and the Columbia Shuttle accident. This rhetorical analysis can help us unpack what happened. Later in the book, for example, I will discuss how eliminations (i.e., simplified rescue plans based on a simulation that predicted Hurricane Katrina) might have (at best) misguided officials or (at worst) demonstrated how government officials did not take the weather projections seriously. Simulation is an emerging and prevalent type of composition worth investigating. If simulation continues to inform national policy and other important cultural events, then we must develop the tools to unpack their meaning and expose their suppositional underpinnings.

Therefore, the purpose of this book is to examine actual simulations and their byproducts in order to better understand how they operate and achieve meaning. This book will extend the growing body of work about computer simulations in technical communication studies and the rhetoric of science. Essentially, it will follow two main research trails: (1) How *do* simulations operate (if not by deception)? Upon what type of reasoning do they hinge? (2) What role does rhetoric play in the construction of computer simulations and their distribution throughout the scientific community?

DEFINITION OF SIMULATIONS: BOTH PRODUCT AND PROCESS

Scientists define computer simulations as products of high-level mathematical equations. Sherman and Craig define them as mathematical models or sets of equations that "describe the possible actions that can take place in the virtual world" (404). Computer simulations create a more complex scenario than sheer paper-and-pencil mathematical equations can because they can handle more involved calculations. The term "simulation" also refers to animations and other illustrations produced by the complex computations (Gould, Tobochnik, and Christian 4). Computer simulations churn out long strings of numbers, to which programmers then assign computer graphical components (such as color and pixel position) in order to generate a visual representation of the data. Computer scientists run multiple variables through the models that they create. Each time they do, they create new sets of data. Each new set translates into different-looking visual components, thereby creating a different picture, be it a frame of an animation or a peak of a line chart. Scientists consider these illustrations another part of the simulation, in addition to the computational model.

Although many scholars call simulations "models," technically speaking, simulations go a step further than mere models. As Morgan and Morrison describe it, simulations capture a model's behaviors (29). While scientists use numerical models to help establish parameters of a simulation, the models in and of themselves represent static features. Simulations, on the other hand, put models through numerous conditions, variables and states. They put models into motion. Models map closely to an actual object, but they differ by scale. Furthermore, simulations have forth-telling and foretelling capacity. Theories define possible states of an object of study, and models confirm or contest portions of those mathematical theories. In contrast, simulations represent actual phenomena by presenting multiple states and the interworking of several theories and models (Winsberg "Simulations, Models, and Theories," S449). Simulations might start with abstract theories and models, but they aim to show something actual (Winsberg "Simulations, Models, and Theories," S449). Since the actual object is the goal, a simulation scientist cannot solely depend on mathematical principles to construct the simulation; he or she might do so when building a model or theory. Simulations are a bridge between two worlds: speculation and calculation, theory and experimentation. Simulations aren't mere theory because, per Winsberg, "theories do not have by themselves the power necessary to represent real, local states of affairs" ("Simulated Experiments," 119). By definition, theories preclude representing local and actual events. Theories and models abstract data; simulations apply it.

However, even the most accurate simulation uses a finite scale that cannot replicate nature, which operates on an infinite scale. Casti describes three realms: (1) the real world of observations, (2) the mathematical world of theorems, and (3) the computational world (203). He explains that computations are twice removed from the real world. First, they are translated into mathematical expressions, then into binary ones. Mathematical theories need not end, since numbers extend into infinity. Computational numbers can only extend as far as the hardware's memory and software's programming can handle. In technical terms, mathematics produces partial differential equations, and computers produce finite difference equations.

Programmers base simulation software on theories and laws from hard sciences. Whether or not scientific laws themselves are rhetorical, computers cannot handle those scientific laws as words or purely mathematical equations (Bokeno; McGuire and Melia). Once scientists translate partial differential equations into finite difference equations, they must input data and manipulate software to create a simulation to answer their particular research question. Of course, the simulation does some writing of its own. Computer programs contain main routines that manipulate sub-routines to generate the data. Sub-routines process input. How (and which) sub-routines activate depends upon what values the input generates. In essence, sub-routines interpret information from the main routine and from each other.

However, simulations also involve a degree of manipulating input on the part of the programmer, too. Unlike mathematical models, simulations require a little ingenuity on the programmer's part. Models replicate theories as closely as possible (Winsberg "Simulations, Models, and Theories," S445). Since theories have close mathematical counterparts, models can often closely follow mathematical logic. However, simulations often require their own internal logic, too. They include ad hoc reasoning: making assumptions, substitutions, and following hunches. In a personal interview with me, astrophysicist Anthony Mezzacappa calls this reasoning scientific intuition, or a "human value system that comes in" to judge whether a simulation parameter is unnatural; subjective elements such as "naturalness and beauty and symmetry make good, albeit subjective, guides." Philosopher Eric Winsberg, on the other hand, calls this ad hoc component of the simulation processes "fudging" insofar as simulation scientists have to assume that whatever they omit from their models only has a negligible effect, and whatever elements they add to the model compensate for whatever they neglected or overlooked (Winsberg "Simulation and the Philosophy of Science," 10). These decisions have rhetorical underpinnings, particularly when it comes to reporting simulations to peers in the scientific community. Chapters 4 and 5 of this book will describe the "fudging" in rhetorical terms; the assumptions that comprise the "fudging" actually correspond to rhetorical figures that simulation scientists use to justify their decision making. In this way, mastering rhetoric can help programmers justify and fine-tune their reasoning. Rhetoric can help uncover assumptions implicit in ad hoc modeling.

Simulations are sometimes called computer experiments because, much like conventional lab work, they test hypotheses about physical systems. However, simulations stand apart from physical experiments because a set of established (and almost universal) methods regulate the latter. Philosopher Karl Popper explained that scientific theories must be falsifiable. He argued that scientists must construct theories in such a way that others in the community can re-run, and possibly disprove, the findings. Otherwise, the community cannot verify facts. Scientific knowledge, then, advances by falsifying misconceptions rather than verifying facts. Popper's rule still applies to simulation science; however, unlike traditional science where researchers have their observations to enlist as they go about testing the falsehood of a finding, simulation scientists may not have that luxury. In some cases, simulations are often predictive, so the actual event they represent hasn't happened yet. In other cases, they represent events from the very distant past or remote location, both of which preclude direct observation. Simulation scientists must falsify future hunches or hypotheses about what will happen, rather than what phenomenon is happening now or did happen in the past. In this way, simulation scientists must judge not the present truth but the potential truth of simulated data or findings. According to Winsberg, simulations "often yield sanctioned and reliable new knowledge of systems even when nothing like the stringent conditions [of traditional experiments] . . . are in place" ("Simulated Experiments," 116). Experimenters practice longstanding scientific methodology. Simulation scientists, on the other hand, have (relatively speaking) only just begun reifying their computational procedures. Computational processes still involve the scientific method—namely, making hypotheses that scientists test and discuss. However, centuries of scientific experimentation have proffered a shorthand and protocol that decades of scientific computation have yet to standardize.

In addition, unlike experiments wherein scientists calibrate instruments and physical samples, simulation scientists must test computations for their accuracy (Gould, Tobochnik, and Christian 5). When a problem arises in a lab experiment, scientists look to correct instruments. In simulation, scientists and programmers correct code. Computer simulations not only suffer from human error, but also from typical programming or coding errors. There are three types (Woolfson and Pert 31–32):

> *programming errors* caused when the computer incorrectly compiles the code and the parsing yields a logical but nonetheless incorrect output;
> *numerical errors* caused by round-off mistakes the computer makes because it can only handle large-but-still-finite number of values; and
> *modeling errors* caused by parsing a problem in such a way that it yields an incomplete representation of the phenomena in question.

For the most part, these methods treat errors as a function of the syntax and semantics of the code. The "fix" involves checking the code's accuracy by mathematical standards.

However, the ad hoc nature of simulations opens up additional sources of error beyond computational or mathematical ones. A simulation scientist commits eliminative errors when he or she neglects to include a certain significant feature of a system (Winsberg "Simulation and the Philosophy of Science," 29). A simulation scientist makes ad hoc creative errors when he or she adds artificial elements that improperly substitute for eliminated ones

(Winsberg "Simulation and the Philosophy of Science," 30). He or she makes an error of symmetry when he or she uses an incompatible substitute to replace an eliminated feature or to compensate for other inadequacies in the computational representation of the theories. For example, if his or her computer will only allow creating a limited-sized spatial grid on which to build the simulation (Winsberg "Simulation and the Philosophy of Science," 30). Symmetrical errors occur whenever scientists model the physical systems as either spherical, axi- or non-symmetrical (Mezzacappa, personal interview). A system's symmetry describes its shape and pattern. Scientists assume that spherically symmetrical systems behave the same and have the same components throughout the entire system. No matter which corner of the system you observe, you can assume that every other inch of it behaves the same way. These systems translate into one-dimensional computational models. Scientists presume that axi-symmetrical systems have two sides that behave the same. For the sake of analysis, scientists measure the system along an axis that divides the similar sides down the middle. These systems require two-dimensional simulations. Non-symmetrical systems contain dissimilar component parts. They require simulations mapped onto at least three dimensions. The fewer dimensions, the more reduction and simplification of the system's features, and the less sophisticated the representation (Winsberg "Simulation and the Philosophy of Science," 31). Finally, a simulation scientist can also perpetrate errors when he or she sets the parameters of the simulation. For example, he or she must set contrived increments and initial values that serve as constants as the simulation progresses from time step to time step. Setting improper initial conditions could yield contrived or particular results (Winsberg "Simulation and the Philosophy of Science," 32).

How is it, then, that simulation scientists validate these conceptual errors? A simulation scientist calibrates his or her program with other simulations, analysis or experiments. He or she can run multiple, independent iterations with randomized seed data. "If the two different implementations yield identical results, then it is unlikely that they each contain the same errors producing the same results" (Winsberg "Simulation and the Philosophy of Science," 32). Winsberg also recommends running other simulations—that is, without ad hoc components or with different symmetries—to check for error and discrepancies. Simulation scientists also analyze their simulations by tweaking the duration of time steps and hypothesizing what results the simulations should yield. Winsberg stresses that empirical data (from telescopic observations and lab work) also improve simulations. Regardless what methods he or she uses, a simulation scientist can expect only so much from his or her results. "[O]nce a simulation study has been subjected to epistemic scrutiny, often the best we can hope for is that some general qualitative features of the results will remain sanctionable" (Winsberg "Simulation and the Philosophy of Science," 39). This book posits another method check: simulation scientists might also consider reexamining rhetorical strategies and assumptions underpinning simulation architecture.

Thus far, theorists have concerned themselves primarily with epistemological nature of simulations (Winsberg 2010; Peterson 2012). They examine how simulations produce knowledge differently than traditional experiments, models and theories. In this book, I will extend these efforts. I will show how simulation scientists use particular rhetorical strategies to make and report ad hoc and other simulation choices. Winsberg alludes to such a connection: "[I]t is also extremely helpful to have a good understanding of where the model comes from and why it might work" ("Simulation and the Philosophy of Science," 35). For example, he continues, astrophysicists had to include a mystery value ($\propto$) to represent an anomalous viscosity in their simulations of accretion disks in order to produce numbers that jived with observations and laboratory data. Unfortunately, the value $\propto$ did not have a known physical counterpart or explanation, and they added it "even though they knew that true, molecular viscosity could not possibly be a factor in these highly rarified systems" (Winsberg "Simulation and the Philosophy of Science," 35). Later analytical studies revealed that turbulence caused the $\propto$ viscosity. The revelation allowed researchers to come up with an equation for $\propto$. In this case, the simulation scientist had to add an incorrect element in order to preserve the integrity and internal logic of the simulated model and to construct as authentic and accurate a representation as possible. Once the model proved itself useful in yielding accurate predictions, the "falsehood" of the incorrect element ceased to matter, and, in fact, became essential for generating reliable information about the natural system that the model was supposed to explain (Winsberg "Simulation and the Philosophy of Science," 36). Whereas Winsberg has mathematical or analytical knowledge in mind when he describes the truth or falsehood (the accuracy or inaccuracy) of the data, I argue that rhetorical understanding is equally important in evaluating the quality of simulations.

CHAPTER PREVIEWS

The rest of this book will perform rhetorical analyses to illustrate the importance of rhetoric in constructing a simulation's meaning. Chapter 2 introduces the idea of studying simulations as rhetorical arguments. It uses the case of a simulation of a complex protein to explain how the steps involved in creating simulations hinge upon rhetorical decision making. It describes how simulations derive their meaning from deliberation and negotiation of facts, from relationships between texts, and from situations in which they are written and distributed.

8 *Chapter 1*

Chapter 3 explains what kind of evidence simulations produce. It compares computer simulations to other strategies common to the scientific imagination. The chapter introduces another case—a simulation of bumblebee flight—to distinguish simulations from paradox, thought experiments and metaphors. In terms of inventing scientific ideas, simulations work differently from these three modes. Chapter 3 also introduces and defines virtual evidence.

Chapters 4 and 5 use rhetorical analysis (rather than philosophical or computational analyses) to explain how simulations are meaningful. These two chapters examine the case of a supernova simulation to illustrate the social context that shapes how scientists settle on the meaning of the simulation. Both chapters examine texts surrounding the case of one particular simulation of a supernova. They analyze documentation surrounding the simulation (such as the published manuscript about it, comments embedded in the code that created it, articles that cite it, and coverage of it in the public) using textual and rhetorical analysis to expose the social and discursive processes and practices involved in creating simulations and constructing knowledge from virtual evidence.

Chapter 6 discusses two recent cases of simulations in current events: (1) reports of the simulations predicting climate change and (2) training exercises based upon a simulation that predicted damage to the Gulf Coast caused by a hurricane of Katrina-like proportions. Ultimately, the chapter discusses the rhetorical dynamics underpinning reception of these simulations, and it shows how politics and complexity of arguments underpinning the simulations affect how simulated evidence is perceived and cast by the general public and stakeholders in these high-profile, highly-charged, and highly-contentious social matters.

The conclusion summarizes my analysis and teases out further implications for future work. It recaps how simulations are virtual evidence that reflects relational meaning. It gestures towards identifying generic characteristics of computational scientific articles. It draws conclusions about how the findings of the rhetorical analysis pertain to epistemologies of science and gaming studies. Finally, it suggests further directions for studying the rhetorical dimensions of computational science.

Chapter Two

The Rhetorical Situation
of Simulations

Rhetoric factors in both products and processes of simulations. Rhetorical methods also allow us to examine the circumstances and influences surrounding and impacting both the process and product. Rather than search for meaning in the text itself, rhetoricians locate meaning within the rhetorical situation, where authorial, audience, and social influences come into play.

Lloyd Bitzer introduced a dominant rhetorical theory for handling meaning: the rhetorical situation. Bitzer defines the rhetorical situation as a "complex of persons, events, objects, and relations presenting an actual or potential exigence which can be completely or partially removed if discourse, introduced into the situation, can so constrain human decision or action as to bring about the significant modification of the exigence" (5). Bitzer claims three components comprise the rhetorical situation: an exigence, audience, and constraints (6–7). According to Bitzer, exigence exists in the world. An exigence provides the impetus to say or write something. If nothing happened, no one would have anything to say or debate. The audience and constraints, on the other hand, dictate how to say it. Audiences have expectations about different occasions, and occasions themselves dictate certain styles and approaches. Not every occasion, for example, calls for humor. Bitzer envisions a rhetorical universe wherein components have clear distinctions and roles.

Other scholars have assimilated Bitzer's terms, but disagreed with his conclusions. The major revisions of the rhetorical triangle place the brunt of rhetorical agency on different components of the rhetorical situation. Richard Vatz, for example, disagrees with Bitzer's definition of exigence. Bitzer implies that it exists outside of the text and dictates what to expect from it, but Vatz contends that the writer creates it (158). Judith Jamieson believes texts create constraints themselves. Writers abide by generic rules, and audiences rely on them to shape their expectations and decipher the text (Jamieson 163 and 165). Barbara Biesecker uses deconstruction to complicate notions of audience. "*Difference* obliges us to read rhetorical discourses as processes entailing the discursive production of audiences" (Biesecker 126). That texts communicate both what they exclude and include makes it unfitting if not impossible to pinpoint a concrete intended or invoked audience. Despite the differences among these theorists, they share a core commonality. All of them treat the components of the rhetorical situation similarly—namely, as antagonistic points on what James Kinneavy called the rhetorical triangle. Kinneavy coined the term, which triangulates the rhetorical situation between an encoder, decoder and reality itself. Kinneavy envisioned these components at three opposite points, the text constrained between them. This model implies that the distance and tension between the three constitute and bridle the text. The rhetorical triangle and situation have a lot in common. The rhetorical triangle describes rhetorical dynamics in terms of audiences, contexts and rhetors, which correspond to Bitzer's terms for the situation. (Audiences decode, writers or coders consider generic constraints, and multiple exigences comprise reality.) The triangle incorporates audience responses and reactions, so it surpasses referential or ideational theories of meaning, wherein an author's ideas and word choice determine meaning. It also integrates context (exigence, the world) into describing meaning. I will use terms from the rhetorical situation and triangle to unpack the rhetorical context of computer simulations in science.

CASE STUDY: SIMULATING A COMPLEX PROTEIN

Let's introduce a specific case—a sample simulation from organic chemistry—as an example. Organic chemist Brent Iverson uses simulations in research and pedagogy. He uses them to teach college freshmen and sophomores basic chemical reactions essential to understanding organic chemistry, such as the claisen reaction, which occurs when the enolate ion of the methyl acetate molecule reacts with another such molecule. Iverson's claisen reaction simulation depicts the very moment when the carbon atom of the molecule in question interacts with the other. When they collide in the animation generated by the simulation, the molecule shakes quickly. Iverson didn't solicit the shake from the program. That response was generated by Hyper-Chem, a molecular graphics program for creating graphics and animation. It has both a basic graphic user interface (GUI) interface and scripts that allow users to build and render 3-D visuals and animations of molecular structures.

The simulation demonstrates details that traditional illustrations such as figures and static displays (see figure 2.1) cannot convey. In a personal interview with me, Iverson remembered learning the claisen reaction in his undergraduate education. He recalled knowing two things about it: (1) When they performed it in lab, upon combining the necessary chemicals, the flask would heat up, and (2) the professor called that an exothermic reaction. A confusing illustration accompanied Iverson's lab experience (figure 2.1). He recalled never making the connection between the figure and the sensory data. Static displays cannot adequately express kinetic phenomena such as energy flowing through a system during a collision. The animation captures the motion of molecules; motion represents energy and heat.

However, what the simulation shows (the features of the reaction or the energy distributed during the claisen reaction) has its own imprecise features, too. In a lab, it takes millions of collisions before the molecular exchange and the heat cause the claisen reaction (Iverson). The simulation shows one fateful impact. In addition, the software calculations do not factor in inescapable parameters like solvents (in which the molecules float). Rather, the simulation calculates what happens in a vacuum. Therefore, the numbers generated by the computer to create this simulation are highly suspect compared to those from lab experiments. Moreover, the actual data outputs—such as collision rates—are not meaningful in the lab because of the lack of actual parameters. No lab recreation of this very same reaction can duplicate what the computer animation shows. Figure 2.2 shows a still image from the claisen reaction animation.

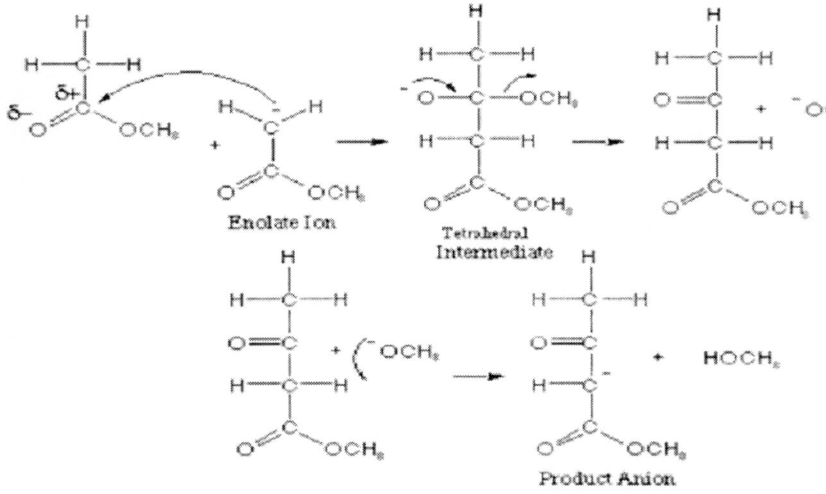

Figure 2.1.   Illustration of the Claisen Reaction. Used by permission of Brent Iverson.

In simulations such as the claisen reaction, multiple stakeholders share rhetorical agency. Therefore, it does not suffice to say that one or the other stakeholder group in the rhetorical situation takes precedence. Who wrote the simulation? Iverson input the data, but the programmers pre-set constraints for the animations. Further, they based the program on theories written by hundreds of scientists and mathematicians. Eliminate any one of the above and the simulation could not exist. What exigence does the simulation embody? It isn't quite the actual claisen reaction, but it bears closer resemblance than any prior illustration, and it is obviously an effective pedagogical aid. Further, if simulations embody nothing more than a virtual representation of the reaction, then what kind of understanding does the audience have of the actual reaction? Who is the audience, for that matter? The class of sopho-

Figure 2.2.   Animation of the Claisen Reaction. Used by permission of Brent Iverson.

mores learning organic chemistry? Or the sub-routines who search out the main routines of the program for data to execute data—the same ones that generate new algorithms from core equations per each new data set?

Iverson's numbers do not convey the claisen reaction, because he has to use inaccurate data for an illustrative animation. The program merely awaits data from users, and sub-routines await data from the main routines. Both are based upon years of research and equations in the field. In this case, Hyper-Chem relies upon years of theory in organic chemistry. The undergraduates see and use the simulation to remember the reaction for tests, but what they see is not actually what they have performed in lab. There is something different about simulations. To limit them to traditional notions of audience, author, and exigence belies their layered and nuanced nature. The text reads, writes, and constrains the meaning as much as the users do. Subtract any of these components and the rhetorical situation is not the same. Because no one agent monopolizes rhetorical agency, boiling meaning down to linguistic markers—that is, connotations, social or otherwise—does not suffice. From what does the claisen reaction derive its meaning? From the actual reaction that the simulation does not, in fact, represent to a tee? Or does it derive meaning from the years of scientific equations underpinning the software? Does the animation gain meaning from how Iverson describes it in class or how students understand and use it? Since it is the case that the animation neglects many of the actual components of claisen reaction, do students derive a partial or counterfeit understanding of the reaction from the simulation?

Biesecker attributes so much agency to the text itself, she comes closest to characterizing the rhetorical situation of simulations; somehow they can create audience and exigence. Without a doubt, considering simulations for their rhetorical merit reveals just how multi-layered the situation is. However, Biesecker's model underplays the role that belief has in rhetorical exchange. According to Biesecker (borrowing from Derrida), *différance* is the operative function that propels communication and interpretation and, thereby, rhetoric. *Différance* makes it possible to understand signs and symbols. Communication amounts to a struggle to represent the object or what Biesecker calls the "'originating' moment" that ultimately is "covered over, or finessed into a unity by writing and speaking. . . . [T]here is invariably a moment in the text 'which harbors the unbalancing of the equation'" (117, 120, and 121). Texts struggle to hold together and keep a unity of terms that exist only by virtue of their opposite. However, to suggest that rhetoric primarily exposes struggle only characterizes one side of the coin.

Communication and interpretation—the core of rhetoric—also enlist consensus and validity. Consider H. P. Grice's maxims of conversation, for example. Grice found that in everyday conversation, communicants rely on one another to abide by cooperative principles of relevance, quality, quantity and manner. Grice's maxims assume that the speaker will try to give only enough of what is necessary and useful in order to understand his or her point. These maxims also apply in discourse communities wherein scholars mean to add to a communal body of knowledge. A sense of truth plays as primary a role in interpretation as *différance*. Bernard Williams makes this point while explaining human language at its core. He describes how language is wired in such a way that, in order to learn it from a very early age, one must assume that utterances are spoken in the context and for the purposes of sharing truth (4). Lies and slippages certainly happen, and negotiating the gray area between truth and falsehood is the stuff of adulthood that separates mature from novice rhetors. However, as Grice's maxims suggest that communication, cooperativeness, clarity, and honesty are the basis of successful and ethical communication. Certainly, the process of rhetoric teeters between doubt and assent. In a classroom setting, students—fledgling scholars that they are—will fluctuate between trusting what they hear and questioning it. However, another important aim of rhetoric is some measure of the agreement or understanding that Williams and Grice describe here. The best rhetoric—the most effective—will acknowledge and account for the teetering and *différance* that theorist suggest are at the core of human communication, and it will strive for the consensus that Williams suggests.

Scientists certainly collaborate toward a shared sense of meaning along these lines. Harold Brown discusses the process of scientific judgment, whereby scientists "gather information, apply whatever rules are available, weigh alternatives and arrive at a judgment . . . discuss our judgment with our peers and reevaluate that judgment on the basis of their recommendations and critiques" (226). Even as the scientific community finds and confirms natural laws, they do so vis-á-vis a process of consensus-building and deliberation. Simulations incorporate peer input at every stage. The information gathered comes from the immediate team manipulating the software. Teams of scientists weigh their options and confer with one another as they make simulations. They also opt to publish findings and let their community of peers reevaluate simulations. Consensus between interlocutors is as important as *différance* to their messages and meaning. Indeed, *différance* provides impetus for rhetorical analysis, but so does a desire for consensus.

In summary, focusing on one or the other point on the rhetorical triangle—sliding from a text-centric focus to an exigence-privileging focus or audience-privileging focus—yields an analysis far too simplistic to account for complex processes such as simulations. In the case of the claisen reaction, the final product (the animation, for example) belies the simulation process that produced it (e.g., that the text itself both reads and writes, that the writer must often disregard reality in order to produce a successful simulation, and so on). Furthermore, single-focused accounts of the rhetorical situation often yield limited theories of meaning. Theories that privilege one or the other point of the rhetorical situation often neglect the fact that meaning involves all of those stakes. For example, empirical studies of computer program comprehension and style often ask computer scientists to read and recreate complex lines of code. By doing so, they imply that readers and/or writers inscribe the meaning of the code, a primarily linguistic—or, at least, syntactic—definition of meaning, rather than a rhetorical one.

RELATIONAL MEANING OF SIMULATIONS

The most important contribution that a rhetorical analysis of computer simulations affords is that it helps render a relational meaning of simulations wherein "language acts principally to make relationships conspicuous" (Cherwitz and Hikins, *Communication and Knowledge*, B1). Rather than simply "point out things in the world," language "creates awareness of relations and their terms" or the "relational conditions among extra-linguistic phenomena" (Cherwitz and Hikins, *Communication and Knowledge*, 81). Cherwitz and Hikins define rhetoric as "processes of discovery" through "differentiating, associating, preserving, evaluating and viewing in perspective, we discover the world we live in" (*Communication and Knowledge*, 110–111). Meaning is created through texts, contexts, cultural clashes or states, and people's ideas. Extra-linguistic phenomena are lived experiences that inform how people make and interpret meaning, but aren't necessarily made explicit in published drafts or other final products. Thus far, studies into the meaning of computer programs limit themselves to the text of program code. Prior research only partially addresses social, rhetorical and other influences, if at all. Rhetorical theories tend to question how other factors—audience, peers, social constraints, and history, to name a few—contribute in making meaning. Therefore, rhetoric can potentially produce a more integrated analysis, one that goes beyond the text itself. In the case of computer simulations, the text and language amount to several components: the code and comments that generate data, and the visuals produced by the computations.

Rhetorical analyses of computer simulations require us to look beyond this text itself and incorporate external factors (such as the decision making that goes into designing and interpreting the simulation program) that influence how simulations are created, used, and perceived. External factors include social factors that influence how people think. Social influences help comprise individuals' perspectives. Ferdinand de Saussure explains how language—with its abstractions and symbolic nature—has the ultimate purpose of communicating what we perceive are realities. We rely on communities and consensus to judge the reality of an object (Saussure 15 and 113). The ideas in an individual's head—and his or her estimation of those ideas—have social value. Even a personal word choice, for example, is not exclusively personal. Rhetors base their decisions on the social value of that word and what it means to those who will hear it. Furthermore, social circumstances themselves enable meaning. Lester Faigley stresses that composition theory must make connections between discourse and social, material, institutional, and cultural structures, particularly when examining writing in non-academic, workplace settings. Faigley recommends a social perspective in research, in which meanings emerge from "relationships to previous texts and the present context" (235). Likewise, simulations call upon prior studies and theories and hypothetical or future eventualities.

Documents—such as articles, animations and other final products—do not contain meaning; they point to its processes, experiences, and stakes. If it is true that language reveals relationships, and relationships enable meaning, then the language does not contain meaning, but rather chronicles it. Meaning is indicative of relationships; it isn't just in the words we use or messages we form, but also our negotiations between other stakeholders. Simulation code "means" or points to relationships between scientific theories, programming languages, past understanding of a phenomena, and expectations about something new (i.e., what we think is happening now, but won't know for sure until other evidence surfaces and forever validates the simulation). Later chapters in this book will analyze a cluster of texts, all of which chronicle some component of the meaning of a simulation. Further, the book will show how simulations also mean social dynamics; they reveal programmers' desires to distinguish themselves from others, for example. Even the simulation code "means" or averages several genres—both mathematical language and narrative comments and citations.

Relational meaning resembles social constructivist notions of meaning insofar as both weigh cultural context heavily in meaning-making. Ludwig Fleck helped define social constructivist thinking when he argued that facts are not absolute, but rather a form of resistance to arbitrariness. "In the field of cognition, the signal of resistance opposing free, arbitrary thinking is called a fact" (94 and 101). Then, theorists like David Bloor and Bruno Latour extended this way of thinking. Bloor advanced the Strong Programme of the sociology of scientific knowledge, which asserted that social dynamics influence all scientific theories, no matter their truth-value. Latour, in turn, helped inspire actor-network theory, which does not differentiate between human and non-human agents in the process of meaning inscription or "all the types of transformations through which an entity becomes materialized into a sign, an archive, a document, a piece of paper, a trace" ("Pandora's Hope," 306). Social constructivist and relational theories of meaning both acknowledge that words derive meaning insofar as they point to lived experiences, relationships, and associations.

While both social constructivist and relational theories foreground the social context and activities that comprise meaning-in-the-making, social constructivism has received criticism from Alan Sokal and other postmodern cultural critics. Sokal voices the sharpest objection; he warns that many social constructivists confuse representations of nature for nature itself. Per Sokal, social influences cannot completely explain the difference between truth and falsehood. Sokal infamously published an article riddled with falsehoods in an academic journal as a means of testing the academic rigor of the peer review process. However, critics overstated the lesson learned from his hoax; many suggested that it demonstrates how scientific facts are true merely by the fact of the scientific community accepting them. Sokal bristled when social constructivists implied that scientific deliberations and the scientific community manufacture laws and truths about the external world via consensus-building. I agree that theories of meaning must not reduce external realities into subjective abstractions.

I will say, however, that social constructivist notions of meaning have yielded some helpful methods for unpacking the multifold dynamics that conspire to make meaning. Recently, Stephen Witte has theorized a constructivist semiotic that neither "depend[s] upon print-linguistic formulations of writing nor approaches that insist on maintaining the 'language before/during thought' assumption," but, instead, assumes a "'both-and' position" (289). Witte bases his semiotics on Charles Saunders Peirce's triad semiotics: "a sign, its object, and its interpretant, this tri-relative influence not being in any way resolvable into actions between pairs" ("Pragmatism," *Essential Peirce*, 2: 411). According to Peirce, action and objects in the world bear upon how we understand signs. Therefore, semiotics cannot exclude world phenomena in and of themselves, nor can it reduce phenomena to mere figments of the viewer's imagination or interpretation.

Peirce envisioned semiotics differently than Saussure, whose semiotics involved two components: the signifier and the signified. Witte enumerates multiple reasons why Peirce's semiotics is better than Sassure's, but I find one particular distinction most important: Peirce's model factors in action and influences that exist in the world in which we live. (For this reason, philosopher Richard Rorty and others have called Peirce a realist.) I believe that Peirce's model embodies the spirit of Faigley's and others' calls for revisions of textual analysis methods. Rather than examine how individuals understand and depict their thoughts about the world, textual analysis can open foci to expose and see how the world (objects and actions therein) influences the thoughts and ideas that we write, and vice versa. For example, in the case of the claisen reaction, when comparing the static model and the simulated animation of the claisen reaction, the animation had more explanatory power to help understand how the actual reaction works. Peirce's notion of linguistic meaning aligns with Cherwitz and Hikins idea of "extra-linguistic" relationships and experiences that comprise meaning. A parent who brings a shopping list to a grocery store calls upon several other texts and reference points that serve as substantiations of his or her lived experience. The list represents ingredients from recipes he or she wants to cook, family preferences for certain dishes or goals for healthier diets, good conversations and good company that he or she anticipates experiencing if he or she chooses the best tasting components, and so forth (Witte 264–265). The customer's past interactions with recipes and the grocery store, his or her present concerns about satisfying the family, and his or her expectations about future meals all informed how he or she made, used, and interpreted the list. In this case, "extra-linguistic phenomena" such as the relationship between the shopper and his or her family and friends, knowledge of eating habits and preferences, and desire to create pleasant dining experiences in the future all mattered to how the shopper read the list. The list itself—comprised of nouns, phrases and scribbles—only partially suggests such connections.

Likewise, I want to reinforce neither extreme. I want to promote neither absolutism nor constructivism, neither positivism nor relativism. Consequently, I'll call upon the term, relational meaning, to underscore that rhetoric is a means of using language for describing reality as a "vast multiplicity of relationships that . . . reveal themselves not in their totality, but by disclosing limited aspects or circumscribed dimensions" (Cherwitz and Hikins "Irreducible Dualism," 239). Relational meaning retains a sense of realism, or, at least, a resistance to equating the world with our perception of it. The conclusion of this book will discuss what data from my analysis of actual simulations can lend the debate still waging between those who see scientific knowledge as mostly social construction (such as Latour, and Gross) and those who say it is as much a product of natural realities as cultural circumstances (such as Sokal, and Cherwitz and Hikins). This book will also use several cases of computer simulations to demonstrate the relational nature of how simulations acquire meaning.

Chapter Three

Simulations and the Scientific Imagination

In chapter 1, we discussed how scientists make seemingly unfounded decisions when simulating phenomena. This ad hoc reasoning—wherein scientists make assumptions, reductions, and other subjective choices for the sake of a simulation's believability and capacity to represent an object—distinguishes simulations from models, theory, and experiments. Simulations differ slightly from models insofar as the former are "rich inferential process[es] and not simply a 'number crunching' technique" (Winsberg "Simulations, Models, and Theories," S442).

This chapter will further distinguish simulations from other scientific genre. In doing so, it will help answer questions about the nature of simulated evidence. First, rather than define simulation as a trope (as do many other rhetorical theories involving simulation), this chapter will place simulation within the family of rhetorical devices that comprise the scientific imagination. Gerald Holton is well-known for ideas on scientific imagination. According to Holton, the scientific imagination incorporates thematic, visual, and analogical thought; it is a rational act, but also a creative process. This chapter will also describe how, up to this point in rhetorical theory, imagination has been insufficiently covered and mistakenly linked to emotion. In this chapter, I argue that rational processes can play a crucial role in imagination, and I discuss the scientific imagination in rational terms. The chapter will establish how ad hoc decisions employ unconventional argumentation besides deduction or induction, and how ad hoc reasoning instantiates virtual rather than other types of evidence.

The chapter will distinguish simulation as a rhetorical device apart from three other rhetorical devices within the scientific imagination: metaphor, paradox, and thought experiments. In order to explain the differences, I will also introduce a new object of study—a simulation of bumble bee flight—to illustrate the rhetorical machinations underpinning ad hoc reasoning. This example will show how rhetorical decisions manifest themselves in the core of simulation technique. I will also show how alternative logic underpins ad hoc reasoning and differentiates simulations from the inferential processes that govern metaphor, paradox, and thought experiments. The line of investigation in this chapter is valuable because it lays groundwork for reading and understanding simulation as a rhetorical act, and it helps establish essential terms for understanding how it is that simulations convey meaning by answer the following questions: What kind of reasoning does simulation entail and what kind of evidence does it embody?

IMAGINATION IN RHETORICAL STUDIES

Unfortunately, rhetorical studies have not sufficiently taken into account the scientific imagination. In fact, James Berlin, James Kinneavy, and others differentiate imagination from scientific investigation and communication. The terms of discussions about imagination are set in a way that excludes scientific imagination. Two points come to mind: Contemporary rhetorical theory (1) maintains a Romantic notion of imagination and (2) conflates non-rational and irrational discourse. In this way, contemporary rhetorical theories articulate imagination in a way that misrepresents the phenomenon.

Usually, rhetoricians define imagination in conflict with rational processes of rhetoric. Imagination escapes reason, therefore rhetoric has difficulty incorporating it. Joshua Gunn argues that rhetorical theory rejects imagination, the essence of invention, because imagination exists outside of consciousness. Rhetorical theory—saturated with deterministic bias—either mishandles the imagination or ignores it altogether because of its "unwillingness to let go of the Cartesian ego, the autonomous humanist subject who claims mastery over the material world in *conscious* thought" rather than embrace "a more contingent and fragmented understanding of individual subjectivity, community, and world" (Gunn 41–42). Gunn describes how extremely rational theories of rhetoric fail to incorporate the components of imagination essential for a thorough discussion of invention. As a remedy, Gunn suggests that new, epistemic theories of rhetoric and other contemporary counters to current traditional rhetoric must give a fuller account of imagination; rhetorical theory must treat "ideology and lesser social forms as having both a mass or political and an individual, psychical existence" (Gunn 53, 55). According to Gunn, even the imagination contains social and ideological imprints. Unfortunately, says Gunn, rhetoric neglects these social, psychical interiors. For other contemporary, traditional critics, the imagination is a product of forces outside of the individual's control and volition.

Current deliberations about imagination's place and role in rhetoric are nothing new. Longstanding debates in schools of thought about rhetoric set the scene for how rhetoricians currently discuss imagination. Gunn cites Richard Kearney's three opposing paradigms of imagination spanning centuries of philosophy and critical theory (Gunn 44). (1) The classical period saw imagination as a mirror that produces mental images of the world. (2) The modern age conflated imagination with human capacity to create meaning. Finally, (3) postmodernism treats the imaginary rather than the imagination. In the former, ideological or subconscious mental influences beyond the individual's control determine his or her consciousness and, therefore limit his or her capacity to direct what he or she imagines.

Gunn cites specific examples from each period of thought about the imagination. Classical and Romantic rhetoricians situated imagination with memory and style. Cicero and Bacon are textbook examples of this; for both, imagination helped invigorate the reasoned arguments that are "in themselves, dreadfully boring" and, thereby more likely move audiences to "moral action" (Gunn 45). It also helps render "representations of a speech for handy recall during performance" (Gunn 45). For Cicero, imagination served as a mirror. For Bacon, imagination breathed creativity into arguments. In the former, imagination duplicates for the audience what the rhetor remembers. In the latter, it regulates a mental picture's degree of vividness.

Contemporary theorists draw from these schools of thought. LeFevre understands invention as "an internal dialogue with an imagined other" (33). LeFevre bridges the distance between the modern and postmodern approach to rhetorical imagination, where forces beyond a rhetor's control govern his or her imaginings. In LeFevre's estimation, the individual rhetor has limited agency. An "imagined other" imposes upon internal thoughts. Also, consider Ernest Bormann's theory of symbolic convergence. Bormann (1972) contends that people use fantasy to cope with reality, and they have dreams about things that don't exist. Bormann was led to the idea that group- or community-centered rhetoric inevitably contains fantasy themes, types and visions, and "that there is a connection between rhetorical visions and community consciousness" (Gunn 48–49). According to Bormann, rhetoric can track group fantasies, themes, ideas and other public vehicles of imagination in order to decode the rhetorical implications therein.

The irony is, both camps—both opponents and proponents of current-traditional attitudes about the imagination—measure imagination by its distance from rationality; for both, imagination escapes reason. For current-traditionalists, imagination has no place in rhetoric because reason governs rhetoric. For the opposition, current-traditional rhetoric can never do justice to the imagination because reason governs rhetoric. Both factions consider the two faculties—reason and imagination—mutually exclusive. The theories summarized above define imagination in contrast to rationality insofar as current traditional approaches obscure imagination, or exclude it altogether.

These and other approaches uphold a Romantic ideal of imagination. Like Wordsworth and many of his contemporaries, current rhetorical theorists characterize imagination as spontaneous, uncontrollable, subliminal, and untraceable. Wordsworth described man's capacity to dream up poetry as the spontaneous overflow of feelings recalled in a moment of calm *(Preface to Lyrical Ballads)*. Carl Jung and his devotees had similar Romantic ideas about imagination (Jackson 354–355). Jung developed active imagination, a technique to "translate the emotions into images—that is to say, to find the images which were concealed in the emotion" (Jung 177). Here, emotions hide the images that comprise the human imagination. These and other theories of imagination and invention imply that creativity is outside of the bounds of reason. They also imply other problematic warrants: (1) that reason cannot fathom the source or machinations of imagination, and (2) that both creativity and imagination are mostly solitary endeavors.

Not only do the aforementioned theories romanticize imagination, they also collapse the difference between irrationality, arationality, and non-rationality. Arational means that "there are no adequate reasons for it and it was taken on the basis of taste, fortune, etc." (Walczak). Non-rational means that "there are adequate reasons for it but either it was not made or an opposite decision was taken" (Walczak). Finally, irrational means that "there are adequate reasons for it but one did not use them and a decision was taken on the basis of inadequate reasons" (Walczak). Arational thought makes no reasonable connections. Non-rational thought disregards them. And irrational thought contradicts them. How mutually exclusive are imagination and scientific investigation? It depends on who is characterizing them. Both current-traditional rhetoric and its opponents imply that imagination is arational in nature—it escapes rationality altogether. I contend that imagination has rational functionality. Even if critics only grant that imagination has non-rational and irrational dimensions, non-rational and irrational communication must enlist the rational rules that they flout. Analyses of non-rational and irrational communication must broach rule-based constraints in order to understand how non-rational and irrational processes break those rules.

Other theorists define the imagination practically. Lev Vygotsky, for example, attempted to revise contemporary working definitions of imagination. Vygotsky describes how theorists typically define imagination as "what is not actually true, what does not correspond to reality, and what, thus, could not have any serious practical significance" (Vygotsky 9–10). Vygotsky corrects the misconception by essentializing imagination as "the basis of all creative activity . . . an important component of absolutely all aspects of cultural life, enabling artistic, scientific, and technical creation alike" (Vygotsky 9–10). Any human endeavor is a product of "creation based on . . .

imagination" (Vygotsky 9–10). For Vygotsky, imagination in science births technology and innovation. Current work in psychology also sheds light on the nature of the imagination. For example, psychological studies indicate that rule-based systems enable and enhance a child's ability to make-believe and play (P. Harris). Such research emphasizes the procedural nature of the imagination.

I, too, want to encourage rhetoricians not to throw out the baby with the bathwater, particularly given current promising investigations into the scientific imagination. Earlier theorists have insisted on too radical a split between reason and imagination. Some critics even go so far as to say that reason plays no role in the imagination. Perhaps theories to date have not properly assessed the relationship between rhetoric and imagination. However, we have not exhausted the need to study how reason factors into imagination in the natural and applied sciences.

THE SCIENTIFIC IMAGINATION

Some theorists in the rhetoric of science do trace how imagination plays a part in the work and research of professional scientists. Philosophers and historians have laid the groundwork for such investigations into the scientific imagination. At the forefront, historian Gerald Holton discusses how visual, analogical and thematic imagination sparked scientific revolutions. Holton identifies three means by which scientists imagine: they use visualization, they draw analogies, and their work often centers on personal interest and recurring preoccupations—what Holton calls *themes*.

Concerning visualization, he describes how Galileo's telescope and drawings led Galileo to new insights. Galileo drew ink-wash drawings of what he saw through his telescope, thereby interpreting what he saw and ensuring that others would have the same vantage when they looked through the lens (Holton 186). In that regard, Galileo's drawings served as a better—more focused and representative—indication of the moon's geography than did the telescope trained on the actual object. Regarding analogy, Holton describes how Thomas Young proposed a theory of light by using an analogy: lights are waves in the same way that sounds are waves. Holton elaborates: "Without even stopping to study the details of this surprising and immensely fruitful analogy. . . , we sense the remarkable daring of this transference of meaning" (Holton 190). Understanding how sound works enabled Young to map the same kinds of behaviors onto light. Here, a metaphor opened the door to new understanding of a phenomenon in question. Finally, Holton offers up Albert Einstein as an example of thematic imagination. Holton argues that several a priori themes ordered Einstein's work and influenced his line of reasoning. Before he had the facts or evidence, Einstein's way of

thinking was to prefer "unification of separate parts of the theories of physics, invariance, symmetry, completeness of description, and essentially New-
tonian causality of events rather than fundamental probabilism" (Holton
194). The scientific imagination includes intangibles that advance science
without empirical lock-steps. Visualization, analogy, predilections of
thought, and themes all depend on the subjective perspective of the scientist-
observer. They also help enable scientists to generate new information, *sans*
the scientific method.

Others besides Holton have launched independent studies of the scientific
imagination. Scholars have noted how metaphorical thinking often helps and
sometimes hinders the process of substantiating important scientific theories
(Baake; H. I. Brown; Eisenberg). Philosophers of science have also cataloged
and analyzed uses of paradox and thought experiment in scientific thought (J.
R. Brown; Norton; Panoff and South). Metaphors, paradoxes, and thought
experiments all share features in common with the scientific imagination: (1)
They use imagery and features of narrative; and (2) they often present unlikely or unrealistic combinations of ideas. These characteristics (along with
visualization, analogy, and theme) further comprise the scientific imagination.

On the surface, paradoxes, metaphors, and thought experiments often
seem illogical. All three ask readers to make sense of disparate, unobserved,
or impossible data. Like simulations, the three rhetorical figures also have
internal logic and rational machinations. They "work"—they render meaning—because they can be rationally coded and decoded, whether or not such
cryptography can ever crack every possible interpretation. But what is the
imaginative nature of simulation? Where does it fit within the scientific
imagination? Ad hoc reasoning commmon to simulation has qualities unlike
its counterparts within the scientific imagination. Unlike analogy, it compares similar terms. Unlike paradox, it hinges on possibility rather than impossibility. And, unlike thought experiments, it produces counterfactual rather than contradictory solutions.

THE CASE OF THE BUMBLEBEE FLIGHT SIMULATION

In order to situate simulations within the family of tropes that comprise the
scientific imagination, I will use an example from biology—a simulation of
the flight of a bumblebee. Since the 1930s, a rumor has persisted in physics
that, over dinner, a biologist asked an aerodynamicist about the flight of bees,
and the aerodynamicist did scratch calculations comparing lift to drag (or
resistance). According to the Reynolds number, the bigger the wing, the
bigger the lift. Of course, the aerodynamicist found that there was insufficient lift in a bee's tiny wing to carry the rest of its comparably huge body.

What makes this paradoxical? The truth of the matter is that bees do fly, despite the incorrect calculations churned out by the Reynolds equations.

The discrepancy propelled years of research in aerodynamics. In response, a team at Cornell led by Z. Jane Wang simulated bee flight. As Wang describes, the discrepancy only underscores shortcomings in the science and theories used to describe a bee's flight. Wang's team found that bee and dragonfly wings rotate, double flap, and perform other methods that increase the wing width and lift. As Wang explains, "The old bumblebee myth simply reflected our poor understanding of unsteady viscous fluid dynamics" (Segelken).

Her team and others have gained a clearer understanding by using a computer simulation to replicate knowledge that other teams have gained through new technologies in observational data (i.e., painting bee wings with digitally-sensitive chemicals and videotaping those bees in flight and other techniques). Observational data have drawbacks, however. For example, the paint affects the flight and, thereby, taints the observed behavior. Wang's team simulates bees in flight. Hers is touted as the first ever proof that bumblebees and similar insects produce sufficient lift to stay above the ground. (See figure 3.1.) Here, the intuitive nature of the simulation—that it confirms what the eye can see, despite what theories predict—lend the simulation more validity than theories. But the simulation required Wang to develop new computational tools—or "tricks" as she called them (Segelken).

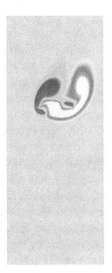

Figure 3.1. Vorticity Field around a Flapping Dragonfly Wing. Used by permission of Z. Jane Wang, Cornell University.

Wang simply applies new equations that field has developed to compensate for the paradox.

SIMULATIONS VERSUS METAPHORS

Historian Gerald Holton and rhetoricians Leah Ceccarelli and Peter Dear have studied metaphors, analogies, and thought experiments in science. Many current theories in the rhetoric of science lump simulation together with metaphor. Granted, both simulation and metaphor are forms of representation (*mimesis*) and comparison, but with a difference. Metaphors (and their cousin, analogies) are a type of *mimesis,* insofar as they underscore the resemblance of two different objects or cases. Some critics conflate models with metaphors: "[A] scientific model is a metaphor. The primary system. . . is the first term of a metaphor. The model is the second term" (Ricoeur 276, 278). It is true that both models and metaphors approximate originals. In addition, the rhetoric of science has documented in detail how scientists use metaphors. For example, Black, Arbib, and others claim that scientific revolutions are metaphoric in nature. When scientists stopped imagining atomic particles as billiard balls colliding haphazardly and started thinking of them as organized units with particular structures, it changed biology, chemistry and biology and, ultimately, enabled breakthroughs in genetic studies (Blackie). The analogy helped them reconfigure the science.

However, although metaphors and analogies have strong sway in science, simulations are distinctive on three points: (1) simulations represent objects with mathematical precision, (2) simulations do not compare objects of difference, and (3) both objects of comparison in simulations are relatively unknown. First, metaphors do not attempt to represent objects with the mathematical and computational accuracy of computer simulations. Unlike metaphors, simulations cannot be fashioned off the top of someone's head. Simulations require far more effort. Furthermore, according to Zoltan Kövecses, "metaphor can be characterized with the formula *A IS B,* where the target domain (A) is comprehended through a source domain (B)" (29). We interpret metaphors by teasing out the relationship between the two terms: "which source domains apply to a particular target" and "which target domains does a particular source apply to?" (Kövecses 118). Aside from the fact that billiard balls are made of atoms, the two have little in common, on the surface. Billiard balls aren't microscopic and atoms aren't played for leisure. However, the difference between the two makes for a fruitful comparison. In fact, the difference enables lessons to be learned about the atom, which, in this case, is the less familiar object. This is not the case in simulations. If scientists had simulated atomic particles, they would have expected the simulation to approximate the particles more closely than the billiard ball comparison.

Simulations do more than simply give an idea about an original; they replicate it. They come close to being it. Later, I'll discuss how simulations contain the *virtue* or *essence* of the object it represents. The target and source domain of a metaphor relate in terms of functional similarities, but need not possess *essentially* the same inherent characteristics. However, a simulation and its subject matter do have *essentially* the same definitive features.

Further, simulations can originate from unintuitive or unpredictable subroutines and internal logic. Metaphors, on the other hand, often hinge upon our being able to understand and predict the behavior of the common term. In chapter 1, we discussed how scientists often resort to using simulations when they cannot replicate their object of study in the lab or when extreme heat, distances, or other conditions prevent seeing the object in person. However, when scientists compared the particles to billiard balls, they meant to associate how both sets function and how they're organized. They did so because they had knowledge about how billiard balls collide. Moreover, they expected their audience to know more about pool balls than the nature of atoms in order to help the former explain the latter. Someone who reads a metaphor can use deduction to unpack its meaning. He or she can compare the general case (something familiar) to the specific case (something unfamiliar) and use what he or she knows about the general case to make claims about the specific case. Simulations do not call for the same understanding of models, since both the general and specific cases might be a mystery or unknown to the scientist.

Simulations set out to capture the essence of an object. They do more than simply give an idea about an original. They attempt to replicate it. In the case of the simulation of bumblebee flight, Wang's team expected the digital versions of to operate, for the most part, as it would in nature. The digital wing flaps virtually correlated to actual ones. While actual wing flaps don't have the same color or size of the virtual one, the scientific assumption is that the actual wings (the source) and the virtual ones (the target) would function almost identically. The simulation does not seek to capture the "gist" of flight, but rather, it seeks to represent actual flight as closely as possible. Going into the simulation, Wang's team knew equally as much about bumblebee flight as they would have going into nature for direct observation. Until they completed the simulation, Wang's team wasn't sure that their theory of flight that challenged the Reynolds number was accurate. They used the simulation as a process of elimination.

SIMULATIONS VERSUS THOUGHT EXPERIMENTS

Simulations also share similarities with thought experiments, another type of *mimesis* that represents theories for the sake of refining or abandoning them.

Scholars disagree about what kind of comprehension thought experiments produce. For example, Roy Sorenson questions whether or not thought experiments test verisimilar rather than essential or actual properties of a phenomenon. Alisa Bokulich adds that thought experiments can only test the non-empirical aspects of a theory, such as consistency and explanatory power. The difference between thought experiments and simulations hinges on these terms.

Thought experiments resemble simulations in many ways. First, both can come to significant conclusions regardless of the validity of their premises. Both also can involve a strong visual component. Both simulations and thought experiments can help audiences arrive at a new understanding of a topic. Often their conclusions alter the course of whatever science or discipline they question. And both simulations and thought experiments also have real aims, but imaginative methods. Di Paolo, Noble, and Bullock describe how biologists Hinton and Nowlan used an evolutionary simulation scenario to demonstrate the Baldwin effect. Hinton and Nowlan's simulation was similar to a thought experiment insofar as both demand "a reorganization of an existing theoretical framework" (Di Paolo, Noble, and Bullock 6). Both thought experiments and simulations can help reorganize existing data and, thereby, facilitate new knowledge. Both rhetorical devices also rely heavily on visualization. Dietrich explains how thought experiments let scientists "use imagination freely as a methodological means to invent new hypotheses" and use it to visualize "properties that could not be grasped in reality but only contemplated in the mind" (316). Computer simulations abide by a similar heuristic principle; they make sensible and comprehensible something that does not really exist. Thought experiments and simulations have potential to shake up and challenge established theories by inviting experts to see things differently.

However, simulations differ from two types of thought experiments, as defined by J. R. Brown. In particular, Brown classifies two super-categories of thought experiments: destructive, which argue directly against a particular theory (J. R. Brown 34) and constructive, which build up an alternative (J. R. Brown 36). Destructive thought experiments "do their job in a *reductio ad absurdum* manner by destroying their targets" (J. R. Brown 76). Constructive thought experiments are either meditative, which "facilitate a conclusion drawn from specific, well-articulated theory" (J. R. Brown 36); conjectural, which "hypothesize a theory to explain [some] phenomenon" (J. R. Brown 40); or direct, which end rather than begin with a well-articulated theory working from unproblematic phenomena (J. R. Brown 41).

Unlike destructive thought experiments, simulations need not employ *reductio ad absurdum*. They usually test hypotheses in earnest, rather than with suspicion. *Reductio ad absurdum* reduces propositions to impossible ends. Simulations, on the other hand, present modal scenarios or possible worlds.

By definition, simulations construct a state of affairs that *could be* the case. Premises of thought experiments needn't conform to actual or possible objects. For example, Einstein imagined himself running to catch up with light beam. "If I pursue a beam of light with the velocity c (velocity of light in a vacuum), I should observe such a beam of light as spatially oscillatory electromagnetic field at rest. However, there seems to be no such thing, whether on the basis of experience or according to Maxwell's equations" (Einstein "Autobiographical Notes," 53). Einstein couldn't physically chase a beam of light; and he needn't in order for the thought experiment to give him a breakthrough about time, space, and speed. Computational simulations, however, have actual or possible counterparts. For Wang's team, the digital wing flaps correspond to actual ones and simulated flight aims to recreate the actual act.

Further, while both constructive thought experiments and simulation can work with existing theories, people can understand constructive thought experiments vis-à-vis simple, intuitive inspection. The same cannot be said of simulation. At best, Di Paolo, Noble, and Bullock write, simulations are the equivalent of opaque thought experiments because thought experiments have conclusions "that follows logically and clearly" and explain their "own conclusion and . . . implications," but with simulations "there is no guarantee that what goes on in it is going to be obvious" (Di Paolo, Noble, and Bullock 6). The thought experiment explains itself whereas simulation requires further analysis and explanation. In Einstein's thought experiment, our perspective changes immediately when we consider the idea of running so fast that light stands still; it immediately challenged his peers' understanding of physics. Contrast this elegant visual with that of the simulation and animation that Wang created. It took the team more than "simple inspection" or immediate result. They tested and retested their model. They compared their simulation to high speed video of actual bumblebee flight.

While thought experiments are based on imaginative claims and presuppositions, they engage those imaginary precepts by deduction. And, as in the case of Einstein's thought experiment, the imaginary claims test rational rules—in Einstein's case, the laws of special relativity. Thought experiments ask audiences to take these claims at face value (or in earnest), apply our current understanding to those claims, and deduce something new about them. In this way, audiences deduce conclusions about the thought experiments, working from understood (or *a priori*) assumptions to new ones. On the other hand, the simulation required several steps, checks, and analyses. The researchers could not consider the results as self-evident.

SIMULATIONS VERSUS PARADOXES

Simulations can also be compared to paradoxes. Bertrand Russell explains that, at their root, paradoxes proceed from contradictions. Olin explains that "they present a conflict of reasons" wherein reason tells us that a statement is true, but also it "seems to tell us that the very same statement is utterly absurd" (5). Panoff and South posit the cause of paradoxes. "Many times a paradox results from incomplete or erroneous data. Another possibility is that our understanding of what we observe may be flawed and in need of revision" (Panoff and South par. 2). The team discusses how, four hundred years ago, scientists tested the validity of Galileo's theory about the sun by looking for parallax or the "apparent movement that close objects make against a more distant background when the point of reference is changed"—a phenomenon that was not completely explained for "hundreds of years, after the invention of photographic plates" and after science abandoned the idea that a star's brightness was only determined by its distance "and that the stars are close" (Panoff and South par. 3). The inconsistency between what astronomers saw, what they predicted, and what scientific laws suggested, lingered for years and inspired generations of research. While both simulations and paradoxes can expose inconsistencies with current theories, paradoxes proceed from contradictions, and simulations proceed from possibilities. Paradoxes hinge upon unlikelihood and impossibility, but simulations hinge upon likelihood and possibility.

In terms of our case for discussion in this chapter, the bumblebee paradox is an apparent paradox, because it can immediately be explained away by faulting laws of physics and calls for revisions that eliminate the contradiction. A scientific paradox presses against and challenges current theory by exposing error or illogical conclusions. The example helps us understand the many ways that simulations differ from paradoxes. As a talking point, let us distinguish between the paradox that inspired Wang and the simulation that she produced. In this case, the paradox underscores logical contradictions, whereas the simulation represents or investigates them. The paradox ended in impossibilities, whereas the simulation ends in possibilities. Third, the paradox can only proceed from common, undisputed facts; the simulation, on the other hand, can proceed from uncertainties. In this way, simulations can yield new data beyond what was originally imagined.

Willard Van Orman Quine writes that paradoxes are arguments against their own existence. He describes the paradox of Frederic, the main character in *The Pirates of Penzance* who turns twenty one-years old after having only five birthdays. It seems impossible that a twenty one-year-old could have only celebrated five birthdays until you consider that he was born on February twenty-ninth (a leap year) and his various travels across time zones could have prevented him from actually celebrating his birthday, but could not prevent him from aging a year. This prime example of a paradox illustrates the "sustaining prima facie absurdities by conclusive argument" that exist both in cases "where what is purportedly established is true . . . [and] equally to falsidical ones" (Quine 2–3). Paradoxes needn't always result in falsehood or impossibilities. Quine distinguishes between paradoxes that actually turn out to be true and ones that can never happen. But all paradoxes proceed from absurd (or contradictory) premises. And they must contain, on their face, the components of their own absurdity, or their own undoing. The same cannot be said of simulations. First, *prima facie* analysis seldom yields conclusive results with simulations, which have intricate levels of equations and reasoning that require more than superficial examination to disseminate. Second, on their face, simulations must contain models with reasonable premises.

How can we gather anything about paradoxes without close inspection? We can because, while the premises might contradict each other, they are not beyond our understanding independently. In the case of the twenty one-year-old who celebrated only five birthdays, we know enough about how birthdays usually work to instantly surmise the contradiction. The paradox calls on us to play out the scenario to its logical conclusion. We use familiar facts and reasoning to undo paradoxes. Quine introduces another type of paradox—antimony—that incorporates familiar reasoning, produces a "self-contradiction by accepted ways of reasoning," and establishes that "some tacit and trusted pattern of reasoning must be made explicit and henceforward be avoided or revised" (Quine 5). For example, Quine continues, many words are homological—they are what they say. The word *heterological* contradicts the rule; it means "'not true of self' . . . we can . . . ask if the adjectival phrase 'not true of self' is true of itself. [I]t is if and only if it is not, hence that it is and it is not"—a paradox (Quine 8). This translucent quality of the logical machinations behind paradoxes enables those who use them to immediately question and challenge the widely accepted rules upon which they are based. In most cases, audiences can engage the premises of paradoxes via deduction. Often, after doing so, they arrive at a reasonable conclusion that makes sense logically. Simulations can also cause those who use them to question original assumptions, but not on their faces and not necessarily immediately. Further, simulations needn't proceed from certainties. In fact, simulations can alter known, accepted theories by starting from uncertain (or incorrect) parameters that yield effective results. In the case of the insect simulation, Wang's team's "tricks" and ad hoc decision making culminated in a simulation that explained a conundrum that the paradox underscored.

ALTERNATIVE REASONING IN SIMULATION

This section will examine what kind of reasoning validates simulation processes and products. Here, the bumblebee simulation will help illustrate how simulations epitomize abduction and virtual evidence.

It has been said that simulations actualize hunches or that they are the middle ground between eyewitness observation and abstract theory. They represent strong hunches that scientists have about phenomena they cannot entirely prove with empirical methods. In this way, they demonstrate another mode of invention available to scientists. Invention is one part of the rhetorical cannon; it finds and makes arguments by generating claims and reasons. Reasoning is important in developing arguments. Aristotle categorized three kinds of reasoning—scientific demonstration, dialectic, and rhetoric. All three employ two types of proofs—deduction and induction. Deduction moves from the general case to the specific; induction, from specific to general.

More modern theorists have further defined these two proofs. For example, Kinneavy identifies three kinds of induction: (1) perfect, (2) mathematical, and (3) intuitive (111–112). Perfect induction offers as rigorous an account of nature as possible. Intuitive induction is the least reliable because it cannot be as rigorous. Kinneavy also identifies two similar deductive moves: (1) generic to specific moves that involve classes or sets wherein sub-sets inherit characteristics from super-sets and (2) those of statements or propositions wherein one precedes the other in chronology and influence (116–118). Further, van Veuren identifies these proofs as two among Aristotle's topoi or topics that serve as generic strategies for making arguments (section 3.2.1). Others have identified alternative proofs, such as abduction. Charles Saunders Peirce defined abduction in the late nineteenth century as an alternative to induction and deduction in science: abduction "presents facts in its Premiss which present a similarity to the fact stated in the Conclusion, but which could perfectly well be true without the latter being so" (*Collected Papers,* 2.1.2.96). In abduction, the conclusion follows loosely from the premise, so the rhetor is inclined "toward admitting it as representing a fact" (Peirce *Collected Papers*, 2.1.2.96). The premise in abduction, then, is "an *Icon*" insofar as it points toward the conclusion in ways that sometimes are not directly referential (Peirce *Collected Papers*, 2.1.2.96). Peirce used abduction to explain moments when scientists decide between two equally appealing hypotheses.

Induction, deduction, and abduction would treat the same premises differently. Take, for example, the different conclusions you can draw from different combinations of similar facts. Deduction takes what is true of the general case and applies it to a specific case: If all the shirts from the middle bin are size medium (general) and the shirts in my hand are from the middle bin, then they must be size medium (specific). Induction works backwards from deduction. It generalizes facts about the specific case: The shirts in my hand are from the middle bin (specific) and they're all size medium. So, all the shirts in the middle bin must be size medium (general). Induction and deduction both use either the general rule to prove the specific case, or vice versa. Unlike induction and deduction, abduction infers an explanation, rather than using one of the known cases to prove the other.

Abduction interchanges the conclusion and the minor premise of a traditional syllogism: All of the shirts from the middle bin are size medium (general) and the shirts in my hand are size medium (specific), so they must have come from the middle bin. The abductive syllogism assumes that no other bins with size medium shirts exist, despite the fact that neither of the premises guarantees the assumption. In both induction and deduction, one or the other of the premises warrants making the leap from general to specific case (or vice versa). In both deduction and induction, you know enough about the general or specific rule to make the logical leap and make a claim. The abductive syllogism reads more like conjecture than the earlier ones. Abduction yields logically problematic warrants—warrants that, from a symbolic logical perspective, aren't tautologies. Abductive statements are not necessarily true; they don't have the logical form that resists falsehood.

Much controversy surrounds abduction. Some theorists believe that although abduction exists in scientific inquiry, all steps of abduction qualify as either induction or deduction (Hoffman). Kinneavy goes a step further and claims that abduction draws the line between scientific and other types of discourse. He contrasts science discourse to its counterpart in referential or exploratory discourse, which he claims primarily operates on abductive logic. He claims that abduction works backwards from evidence to hypothesis. He describes how a "back and forth movement of suggestions and checking is typical of the process of abduction"; two pieces of data are explained by the first hypothesis that must be corroborated by two additional pieces of data, which are explained by a second hypothesis that must be corroborated by two more pieces of data, and so on (Kinneavy 143). Ultimately, induction and deduction ground this back and forth movement. In order to move to new general hypotheses, abductors must find at least two specific features and the prospect of a reasonable third.

I argue not only that simulations enlist abduction, but also that abduction produces virtual results. The bumblebee simulation, for example, argues from an actual case—namely, bumblebee flight as it happens in nature. But it produces a virtual conclusion rather than an actual one. The premises used to produce the conclusion work in the virtual world, but appeared to not work in the actual case. Further, the simulation yielded virtual results rather than actual or observed ones. The animations did not look like the high-speed capture video or actual bee flight—it portrayed air waves in bright colors, for example. But the information that the simulation shows did help the team identify the best equations to challenge the Reynold numbers. The same can be said of the claisen reaction animation from chapter 2. The actual molecules don't look like their animated counterparts; but the reaction does actually displace energy in ways that resemble the animation more than an illustration. Simulations often show what's virtually true about the reaction, if not what's actually or potentially true.

THE NATURE OF VIRTUAL EVIDENCE

If simulations do not yield the same kind of information as do metaphors and thought experiments, then what kind *do* they generate? Simulations yield virtual results, which make for a strange sort of evidence. Peirce developed a concept of virtuality as well as abduction. Peirce describes how a virtual X "something, not an X, which has the efficiency (*virtus*) of an X" (*Collected Papers*, 6.372). According to Peirce, virtual X differs from a potential X because the latter is "without actual efficiency" of the thing itself" (*Collected Papers*, 6.372). A virtual thing has the efficiency or *virtus* of the thing without being the thing. *Virtus* also means "worth" or "power." It approaches, but never achieves, actual status. Peirce cites the difference between actual and virtual representation that fueled the American Revolution (*Collected Papers*, 6.372). According to England, the Parliament needn't contain Virginians or any other colonists, since all members virtually represent everyone in the empire. American colonists, however, they did not believe they had actual representation because none of their own spoke on their behalf in the Parliament. Modern critics have debated what constitutes a thing's worth—whether or not virtual things bear resemblance to the original's form (Heim) or function (Levinson). So far as the bumblebee simulation is concerned, it seems the simulation resembled bee flight moreso than the paradox or other scientific observations of bee flight. Simulations are almost true, but not quite.

Simulations are a type of virtual representation of objects. Achinstein discusses four kinds of evidence—epistemic, subjective, veridical, and potential—in his analysis of a nineteenth century physicist named Hertz whose work revolutionized the science in its day, but was later refuted and revised. In what sense is or was the physicist's work accurate? Either the results are strong evidence without exception, or they *were* strong evidence until a new set of information replaced it, or the results were never strong evidence to begin with because they were inevitably replaced (Achinstein 18). Achinstein calls the first instance—where the evidence needn't answer to anything other than its immediate problem—epistemic evidence. In epistemic relationships between evidence and claims, the immediate nature of the claim determines the truth-value of the evidence. Epistemic evidence is true, provided only that the claim is true. For Hertz and his peers, who had nothing but the present to go by, his work was right because everything available pointed them in that direction. This type of evidence corresponds with induction and deduction, wherein the conclusions logically follow from the premises.

In order to determine what evidence simulations yield, then, we need to discuss the other types of evidence. The second scenario—wherein evidence is valid only insofar as its particular chronological constraints—is called subjective evidence, which is "relativized to a specific person or group" (Achinstein 23). There are two sides to this type of evidence; one that accounts for empirical, and the other accounts for historical truth-status of findings. Subjective evidence accounts for the fact that Hertz's work was believable and true, probable and true, until later scientists refuted his findings (Achinstein 23). The final scenario suggests veridical or potential evidence. Given new developments, Hertz's work isn't accurate and never was. Both veridical and potential evidence require claims to be true absolutely (Achinstein 26–27). Veridical, however, requires "good reason to believe" something. Potential evidence requires less strong reason to believe. To differentiate between the two, Achinstein describes a child who doesn't have the measles virus, but who does have spots on his skin that look like measles (27). In a veridical sense, he doesn't show signs of the measles because the virus isn't present. In a potential sense, he does show signs of measles. Whether or not the virus is present, spots have the potential to signal measles. Once doctors fail to find the virus, however, the child no longer potentially has the measles.

Let us extend Achinstein and discuss simulations. Virtual evidence would claim that the boy virtually—or all but—had the measles. Unlike empirical evidence, whether or not this claim is true does not bear upon the truth-value of the evidence. When we claim that the child virtually has the measles, it doesn't matter whether or not the child has the virus. Unlike subjective evidence, this claim isn't relative to a particular time or place. He can virtually have had signs of the measles both before and after we know whether or not he actually has the virus. Unlike veridical evidence, this claim needn't be true absolutely. It doesn't require the good reasons necessary for veridical evidence. It only required the spots to resemble those of measles. It doesn't require as conclusive evidence as the virus itself. We can use virtual evidence to make stronger claims than potential evidence. Potential evidence only speaks to the status of the spots being signs of measles. Take the logical next step—claim that the child potentially has the measles—and the statement becomes an empirical matter. It's true only insofar as doctors haven't found the virus. The same isn't true of virtual evidence. It is reasonable (though perhaps hyperbolic) to take the next logical step and claim that the child *virtually* had the measles, particularly if the spots on his skin prevented him from going to school, left longstanding scars, or caused any other measles-like results. Simulations—such as the claisen reaction and bumblebee flight animations—are virtual evidence because they can retain truth value even when missing components vital to the actual object.

Traditionally, scientific evidence falls under the category of epistemic or veridical evidence. First-hand observations strengthen the validity of the claims that scientists can make about lab experiment findings. Simulations, however, qualify as virtual evidence, because they rely on virtual observations rather than actual ones. Virtual observations yield important data about a problem. The data needn't be conclusive or absolute to lend meaning. Even though virtual evidence does not perfectly replicate the object, it has the capacity to explain the thing itself because it does have the virtue of the thing—the worth and workings of it. Abduction results in virtual observation. Abductive reasoning does not require that the premises have logical validity in order to be useful or to reflect the essence of the real thing. As with virtual evidence, if you apply strictures of tautologies, abductive enthymemes will not pass the test.

I argue that abduction comprises the ad hoc modeling that characterizes simulation. As discussed earlier, the ad hoc process includes simplifications, substitutions, subtractions, additions, and other subjective calls. This ad hoc reasoning underpins the bumblebee simulation and the claisen reaction (in chapter 2) insofar as the initial parameters of the simulations do not perfectly correspond to the initial conditions of the actual phenomena. The idealized constraints do not degrade the evidence that the program produced, however. Ad hoc decisions—and the abductive logic underpinning them—are programmed into simulation routines and sub-routines. Later in the book, we'll also see that simulation scientists also use abduction to draw conclusions about the data that their simulations generate. The virtual bumblebee flight and the virtual claisen reaction are more complex than a simple metaphor or analogy. In order to make the connection from the animation to an actual reaction, we need not argue from dissimilar or separate but parallel cases. Instead, we argue for the likelihood of an event insofar as it was virtually shown to us.

Metaphors, thought experiments, and paradoxes do not produce virtual evidence of the object being represented. None of those rhetorical devices need preserve the essence or virtue of the object of study. Paradoxes problematize the essential features of an object in question, and thought experiments can employ fanciful rather than essential terms. Metaphor, on the other hand, relies upon the virtue of another, better-known object to comprehend a questionable object of study. Metaphors operate on traditional deduction and induction—an unknown species compared to a more familiar one. Thought experiments behave more like potential than virtual evidence, insofar as they cannot speak to the validity of a claim as much as they can the validity (or falsehood) of the conditions for the claim. For example, with Einstein's thought experiment in which he chases a beam of light, the claim hasn't validity until Einstein has the capacity (thanks to a rocket or other technology) to run alongside the beam. Thought experiments do not stand as evidence for the actual claims so much as for the theories behind the claims. However, if the claims are false, they render both the thought experiment and underlying theories false. Unlike thought experiments, simulations can be actually wrong, but virtually useful.

Virtual evidence also embodies an important dynamic of communication called *energeia*. Rhetorical theorists have recovered a promising concept from classical rhetoric that might better describe the rhetorical situation—the notion of *energeia*. Richard Lanham defines the term as "vigor, vividness, energy in expression," first coined by Aristotle to describe the process of making a description vivid or bringing it "before your eyes" (64). According to Lanham, Quintilian also used *energeia* in a context where he is discussing forming visual pictures (64). Sara Newman elaborated on the term *energeia*. She read portions of *Poetics, De Anima,* and *Rhetorica*, and she surmised that Aristotle tended to place a median or middle ground in most of the categories that he classified. He used "bringing before the eyes" to explain how an idea strives for middle ground. We get a clearer sense of the term meaning—or bringing something to a mean—from this notion.

According to Aristotle, says Newman, proper making of meaning (e.g., perception and transmittance of an idea) happens when all the necessary conditions are met and all the contraries and evidences balanced. Newman also distinguishes *energeia* from *kinesis*. Unlike *kinesis* (e.g., dynamics that move toward completion), *energeia* is "the sort of thing which is perfected or completed in the very instance of its being enacted" (13). *Energeia* accurately describes the aim of rhetorical analysis. It is the capacity of rhetoric or communication to enable concepts and ideas to "come to our senses" as, in the case of simulated science, unobserved phenomena are brought "before our eyes." *Energeia* also most accurately describes simulations, because it connotes meaning *in media res* or in the middle of the action. A simulation is a site of *energeia*, if we understand it as an ongoing process of making transparent relationships between evidence, decision making, and consensus building. Persuasion is the end of that process; moments of achieved meaning (or middle ground) comprise persuasion. This is the best way to describe what makes the bumblebee simulations so convincing. The animation brought the meaning of the reaction before our eyes. Thanks to the animation, the reaction brings to the senses what a paradox made unfamiliar and absurd.

A message can achieve *energeia* in process—as it is being carried out or as it is being fully worked out. Jonathan Beere describes how *energeia* is both the "exercise of capacity to do something" and the "doing of it itself"

(161). Aristotle engages notions of *energeia* and teleology (or the fulfillment of something together. Representations with *energeia* are in the process of "carrying out or fulfilling their function (ergon), and hence, they are in fulfillment" (Beere 162). Completion, then, is not the sense of telos that *energeia* reaches. Rather, it achieves perfection or completion as it fulfills its function. *Energeia* "carries out the task towards which it is intrinsically directed" (Beere 165). *Energeia* suggests that evidence can have explanatory and persuasive power while or before it is being proven—in the middle of action. It can be perfected (or reach its end of serving to inform or persuade) while it is being settled or enacted. *Energeia* implies that a message can be fulfilled without reaching its telos, which means end or resolution. In the case of computer simulations and simulated science, the potency of simulation is reached without its fulfillment or its proof of a one-to-one correlation with an actual event. It differs somewhat from the rhetorical triangle model, in which the message that the rhetor delivers balances or harmonizes the tensions between the points or axes; the target of the message is to complete, satisfy and resolve the tension. In this way, the rhetorical triangle encloses the message. *Energeia*, on the other hand, implies an open-ended situation and an unresolved message, but a message that still has potency to persuade.

In conclusion, unlike paradoxes, thought experiments, and metaphors, simulations hinge upon abductive reasoning. Observers can often use simple deduction to reveal discrepancies of paradoxes. Thought experiments draw hypotheticals and hypotheses out to their logical conclusions. And metaphors describe things by using deductive reasoning to compare unlike objects to better known objects. Simulations, on the other hand, build arguments from conjectural and contingent assumptions and, thus, produce something virtual (very, very close but *not quite*). On the other hand, the simulation's observable success in reproducing an object also enables it to produce something virtual. Understanding virtual evidence in this way helps correct misunderstandings about virtuality, which often suggest that a simulation—the virtual thing—is indistinguishable from the real thing.

Chapter Four

Rhetorical Strategies of Simulated Evidence

Rather than examine simulation from a computational or epistemological perspective, the prior chapter compared simulation to other modes of the scientific imagination, and it concluded that simulation hinges on abductive reasoning and produces virtual evidence.

This chapter will demonstrate how ad hoc deliberations are not disingenuous or purely semantic, but rhetorical and necessary for computer simulation methodology. In this chapter, I'll discuss how rhetorical strategies underpin the construction of virtual evidence and its reception and use in the scientific community. I will analyze drafts, code, citations, and e-mail correspondence of a pivotal study in astrophysics, which introduced new findings recently cited over 350 times in other articles, presentations, and research. This chapter explains how the writing process helps simulation scientists interpret and direct their work. In particular, counterargument, narrative, abduction, *deliberatio, compensatio,* and *aetiologia* underpin simulation scientists' assumptions. They help simulation scientists set levels of complexity and make other ad hoc decisions. Ultimately, my findings also contribute to a body of work that explores how the writing process itself helps mold science along with observations and facts.

THE CASE OF THE SUPERNOVA SIMULATION

Let's examine a simulation project performed by astrophysicists and computer scientists at the Oak Ridge National Lab (ORNL) in Tennessee. Led by astrophysicist Anthony Mezzacappa, the Terascale Supernova Initiative (TSI) team investigates the mechanisms behind supernovae explosions.

Supernovae mark the final stage of a star's evolution. A star begins as a protostar (clouds of gas and dust). Gasses and other elements pulsate through collapsing stars. Gravity pulls the floating particles together and heats up the cluster (a.k.a. accretion), but not at uniform rates or pressures (Herant et al. "Neutrinos"). The inconsistency precipitates an explosion, which has the potential to synthesize chemical elements in the Periodic Table. Knowing more about the explosions has helped explain the Big Bang theory and, in turn, the formation of life on Earth. Since supernova research might help scientists determine how certain elements came into existence, the U.S. Department of Energy (DOE) has interest in it for developing new energy sources and better means of purifying old resources.

The assumption has been that gas waves dissipate and energy dwindles as a star dies. Therefore, some other force or element must provoke the huge explosion. Neutrinos (neutral subatomic particles) within the star's core seemed the most likely culprit (Herant, Colgate, Benz, and Fryer). The TSI team's groundbreaking work, however, revealed that small gaseous perturbations (accretion shocks) in the star's core can accelerate the waves and possibly create enough energy to fuel the explosion. Their simulation proposes an alternative means besides neutrinos for the supernova to generate energy.

The TSI team ultimately published their work—entitled "Stability of Standing Accretion Shocks with an Eye Toward Core Collapse Supernovae" (which I'll call the SAS article, draft, or simulation from this point on)—in the *Astrophysical Journal* in 2003 (Blondin, Mezzacappa, and DeMarino). John Blondin (Department of Physics, North Carolina State University, Raleigh, NC) was the lead author, and Anthony Mezzacappa (PhysicsDivision, Oak Ridge National Laboratory, Oak Ridge, TN) and Christine De Marino (Department of Physics, North Carolina State University, Raleigh, NC) coauthored the article. This chapter examines the publication, drafts, computer code, and correspondence from the authors to each other, all recorded in e-mail transmissions.

Theoretical sciences often depend upon simulations rather than laboratory experiments as primary evidence to explain phenomena. The astrophysics article under examination in this study proposes a new mechanism for understanding how supernovae explode. The case study is also important insofar as it helps to situate ad hoc modeling as a legitimate mode of rhetoric and argumentation.

The team ran multiple one- and two-dimensional simulations to examine the stability of standing, accretion shocks within core-collapsing supernova. The paper contains six sections, including the introduction and conclusion. Instead of one methods section, it distinguishes between idealized (mathematical) and numerical (computational) methods used to build the simulation. The section on idealized models tells which longstanding equations taken from mathematics and physics the team used. The numerical section translates those theoretical equations into computational parameters. The SAS article also divides explanations of the results into two different sections titled "Stable Flow in One Dimension" and "Instability in Two Dimensions." The team used one-dimensional simulation results as a baseline to verify that their simulated supernova produced the standing accretion shocks commonly assumed to exist at their core. The team could then compare the instabilities that their 2-D simulations produced to the stable 1-D results.

The SAS article reports on what makes a star unstable to the point where it becomes a supernova. The TSI team suspected that accretion shocks—growing, concentric waves of motion in the gases and particles—in the core of the star play a big part. In order to prove it, the team had to simulate the supernova on multiple levels and at multiple stages. They had to replicate a star, set its parameters for stability, and reproduce the gases in space that interact with it. Complexity abounds in an actual star explosion. However, TSI's simulation only captured the essence of the complexity, not its entirety. (Refer to the discussion of simulation *mimesis* and the definition of virtual evidence in chapters 2 and 3.) The SAS simulation contains aimplifications, substitutions, subtractions, additions, and other choices on the part of the TSI team. The team could not include every realistic detail; they had to make ad hoc additions, eliminations, and initial values. Rhetoric can help explain the ad hoc decisions that characterize simulation (Winsberg "Simulations, Models, and Theories," S445).

For example, Blondin, Mezzacappa, and DeMarino start with what it calls an idealized model of the shock. It begins with a virtual star assembled vis-à-vis multiple assumptions and shortcuts. When simulating the gas, the team assumes "that the infalling gas has had time to accelerate to free fall and that this velocity is highly supersonic" ("Stability of Standing Accretion Shocks," 971). They also "assume that radiative losses are negligible . . . and the gas is isentropic" ("Stability of Standing Accretion Shocks," 972). The team set these parameters so that they could "separate effects due to convection from other aspects of the multidimensional fluid flow" ("Stability of Standing Accretion Shocks," 972). Likewise, when setting stability levels for the star, the team "must impose an inner boundary . . . [that] allows a mass and energy flux off the numerical grid in order to maintain a steady state" ("Stability of Standing Accretion Shocks" 973). Even though supernova without the team's particular values for the inner boundary might exist in nature, the boundaries make sense in terms of the internal logic of the simulation they're making. These idealized settings are imperative, given what the team wants to show. These constraints border on conjecture insofar as they do not correspond to any actual star. Rather, they posit a virtual one (as defined in chapter 3). In doing so, the team employs inventive methods besides metaphor and analogy when they construct this virtual star. They aren't arguing from dissimilar or separate-but-parallel cases. Instead, they're arguing for the likelihood of an event by virtually showing it to us. As they make the necessary assumptions, the vivid picture of their idealized supernova comes into focus.

In order to track the rhetorical underpinnings of the SAS article, I reviewed multiple prior drafts, the published article, e-mail correspondence, and some of the computer program used to help create the simulation itself. I requested and received the primary sources from Blondin and Mezzacappa. The team volunteered drafts of code and articles for my review. They also sent electronic files with the e-mail conversations about the drafts. Furthermore, the team directed me to the Fortran code upon which they based their simulation—Virginia Hydrodynamics PPMLR 1 (VH-1). I compared two different versions of the computer code partially responsible for creating the simulation—one available online since 1991 and the other since 1998. I analyzed the differences between how the two versions handled comments and semantic cues within the code. Comments are natural-language guideposts that programmers write within code to explain a sub-routine's function; computer scientists have yet to reach consensus on when and how to use them. The Human Subjects and Institutional Review Board at the University of Texas at Austin reviewed the study and deemed it exempt on the grounds that it was a study of existing data and documents.

Equipped with this information, I went out and found articles that cited their work. I also Googled the team and collected presentation slides, abstracts, annual reports, newsletter articles and other secondary documents about the SAS project. To date, the article has been cited in over 300 other articles. Blondin and Mezzacappa gave me drafts of the SAS article embedded in e-mail correspondence dated from February 2002 to October 2002, during which time the team not only drafted paper editions with each other, but also had the paper vetted by journal reviewers. The e-mail correspondence also contained communication between the authors and the reviewer. The e-mail correspondence did not include the very first draft of the SAS article, and it references conversations that the team had about the paper on the telephone and in person. Obviously, my research cannot account for the decisions and thought processes that transpired during those unrecorded conversations (i.e., phone conversations and face-to-face meetings). However, the e-mails and drafts do reveal volumes about how rhetoric plays a part in constructing the story about the process and significance of the simulation. The e-mails record many stages of writing: (1) a first e-mail exchange about the original draft (absent from the e-mails); (2) a first recorded draft containing comments and questions that Mezzacappa interjects (set off with asterisks); (3) a second e-mail exchange regarding the prior draft; (4) a second recorded draft with changes; (5) a third e-mail exchange with new changes; (6) a third recorded draft with editions; (7) a fourth e-mail exchange between researchers and the peer reviewer assigned by the journal editors; and (8) a fourth draft with changes.

I use rhetorical analysis to examine the drafts. Jeanne Fahnestock, Leah Ceccarelli, and others have used similar analyses, which call for close reading of primary materials to identify classical rhetorical figures and devices of argumentation in expository writing and speech. They have argued that figures of speech are essential in science for explaining key scientific concepts and theories. In this way, rhetoric serves not only as a means by which to persuade the scientific community to support one theory over another, but also as a foundational tool in the production of scientific knowledge itself.

Chronologically speaking, the first e-mail exchange and draft establish that the researchers have rhetorical awareness of their writing. The simulation scientists are aware that their work could potentially contribute to the field. Even as early as this first e-mail exchange, the team ponders key features of the report (titles, abstracts) that could miscommunicate the point of their research. These comments culminate in the first recorded draft. The second e-mail exchange and draft find the researchers developing their conclusions. They realize and frame the full significance of their work *in media res* or in the middle of the action, as opposed to *a priori* or *a posteriori* (beforehand or afterwards). The third e-mail exchange and recorded draft contained minimal changes that further demonstrate the researchers' rhetorical awareness of article features (such as format changes to captions and added citations). In the final e-mail exchange and drafts, researchers use the arguments they've constructed to defend their work to a peer reviewer's protestations. Ultimately, once the journal accepts and publishes the report, other teams of researchers use the published conclusions to (1) construct a research territory into which to interject new findings and (2) explain peculiarities in their own research. Chapter 5 discusses the citations further.

For the purposes of discussion, I'll combine changes to the second and third drafts, because those made to the third draft are minimal, similar to those from the first to the second drafts and prior to journal reviewer comments. I'll compare these to the final draft, published in the *Astrophysical Journal* on February 20, 2003. I will also discuss what peripheral documents (such as newsletter articles, meeting minutes and annual reports, correspondence and works that cite the SAS article) reveal about TSI's work.

PHYSICAL EVIDENCE IMPACTING THE SIMULATION

In the SAS simulation and article, Blondin, Mezzacappa, and DeMarino use whatever physical evidence possible to construct and report their supernova models.

Table 4.1.   TSI Timeline

2001

National Energy Research Scientific Computing (NERSC) publishes their annual report, wherein Mezzacappa explains TSI's significance

October
Department of Energy (DOE) Research Newsletter features Mezzacappa interview discussing TSI significance.

2002

NERSC publishes its annual report that briefly summarizes TSI's research and mentions Mezzacappa and Blondin as TSI investigators. Mezzacappa interview featured in ORNL Review, the lab's newsletter.

January
Blondin and Mezzacappa present their poster, "Sources of Turbulence in Core Collapse Supernovae," at the AAS 199th Meeting in Washington, D.C.

June 24
Blondin sends Mezzacappa a revised draft complete with changes addressing Mezzacappa's questions and concerns.

July
Excited about the new draft and what it potentially suggests, Mezzacappa e-mails Blondin to discuss the implications. Mezzacappa sends another revised draft of the article—with new, improved conclusion and discussion—to Blondin.

Mezzacappa gives a final stamp of approval to the revised draft.

August 27
The Astrophysics Journal (APJ) editors send journal referee's reports to Blondin.

September
Blondin and Mezzacappa work through and send a response to referee's questions.

Mezzacappa sends Blondin another revised copy of the article, one in which he rewords and removes some references.

Mezzacappa sends Blondin a reworded sentence for the conclusion.

October

APJ editors send word to Blondin that the revised paper is accepted for publications.

TSI team presents findings at a conference.

2003

February
The article appears in the APJ.

March
Blondin and Mezzacappa present their poster, "The Inherent Asymmetry of Core Collapse Supernovae," at the High Energy Astrophysics Division (HEAD) 2003 Meeting.

Mezzacappa reports TSI progress at the Advanced Scientific Computing Advisory Committee Meeting.

2003 to Present
TSI research gets mentioned in 19 other articles in astrophysics, physics and other sciences.

2004

Mezzacappa and TSI featured in ORNL Review newsletter.

Unfortunately, the TSI team could not use a wealth of direct observation to confirm this theory. Star explosions happen at extreme heats, pressures, and distances; they are inaccessible, save for residual evidence at micro- and macroscopic levels (Mezzacappa). In spite of these odds, astrophysics isn't altogether without physical evidence, albeit indirect. For example, at the macroscopic level, scientists look for evidence that supernovae leave in our atmosphere. Supernovae emit photons across the optic spectrum, and some telescopes can measure gamma rays, x-rays and the like. Supernovae also emit neutrinos that reach this atmosphere. Next-generation neutrino detectors yield some information about supernovae. The evidence is physical, but scarce. To put the scale of the information into perspective, consider the fact that the Kamiokande—the neutrino detector cited in the work awarded the Nobel Prize in Physics in 2002—captured little over a dozen of the estimated trillions of neutrinos emitted from the supernova in question ("Solving").

Nuclear physics also comes into play when supernovae explode. During a personal interview, Mezzacappa described how microscopic observation helps explain astrophysics. Per Mezzacappa, simulations are validated through "astronomical observation," "microphysics—neutrinos interacting with a nucleus," and "nuclear physics terrestrial experiments to make measurements of such interactions." Mezzacappa also warned that "while it is reasonable to assume that the elements behave in supernova or in space as they do in the lab, the assumption is not conclusive." He described how physical evidence only provides a "fingerprint" of supernova explosions.

Mark Wolf discusses how other instruments of modern science often do not observe phenomena in the traditional sense. More and more, these technologies use means other than direct observation to evidence or witness physical phenomena. They use electron microscopes, which are extensions of the "optical microscope and the camera, using a focused beam of electrons rather than light to create an image" (Wolf 272). They use scanning tunneling microscopes, or "computer imaging [that] produces an image by scanning a very small area with a fine, microscopic probe that measures surface charge" (Wolf 272). These tools create visual images that are "not an image in the traditional photographic sense, because traditional photography is impossible on a subatomic scale" (Wolf 278). Instead, they create trace images of motion waves, imprints, and shadows. Computer simulations continue this trend toward abstraction in scientific evidence that Wolf describes. Observation has become less and less direct. Likewise, simulations of natural phenomena are not the phenomena themselves. However, despite the abstract nature of those observations, simulation scientists do draw upon existing physical observations. Ultimately, an astrophysicist wants his or her simulation to make sense in light of what traditional experimental scientists observe, no matter how trace the evidence (Mezzacappa). Therefore, he or she will use even obscure observations from nature and from physical experiments to design and read simulations.

Since supernovae explode light years away, the TSI team had no immediate expectation they'd be able to verify their simulation first-hand. Mostly, their terascale supernova simulations relied upon multiple and deep layers of applied mathematics and computer science to give them a concrete idea of an explosion's mechanism. TSI designed a supernova simulation software built on the following types of equations: (1) equations that represent governing supernova physics, (2) those that govern and integrate physics equations into the software, (3) those that monitor code performance, (4) those that manage and analyze applications of simulation data, and (5) those that develop the data representations and render visuals that help the team understand the data (TSI).

Blondin, Mezzacappa, and DeMarino base the behavior of an unknown (the waves of gas in the supernova) upon what is already known from laboratory physics about the behavior of wave particles—namely, how waves of gas behave under certain levels of pressure, heat, and rotation. The team cites physical evidence throughout the published manuscript and in their response to reviewers. They include the following in an e-mail response to their manuscript's peer review: "We presented Figure 7 to illustrate the feedback between vorticity generated at the shock and pressure waves generated from the turbulent flow at small radii fed by the vorticity." Vorticity is a foundational concept in fluid dynamics. Wave behaviors have been tested and well-established with laboratory experiments. Blondin, Mezzacappa, and DeMarino use observable evidences to help support their claims about an unobservable event—namely, supernova mechanisms. While these facts in and of themselves are insufficient to prove the new mechanism, and while obtaining more physical evidence and direct observation of supernova explosions is impossible with current technology, using commonly shared sets of positions about physical phenomena does help make their case more convincing.

GENERIC CONVENTIONS IMPACTING THE ARTICLE

The published version of the SAS report reveals a small portion of the rhetorical deliberations that went into the simulation.

Many studies have concluded that scientific articles do retell valuable goings-on in the laboratory and during observational studies (Bazerman; Gross). Many agree that scientists consider research articles one of the primary and most orthodox means by which to present new findings and information to the scholarly community (Swales; Rowland). From the very first research article, the purpose was to testify to others of the validity of findings. Science historian Steven Shapin explains how some of the very first reports in physics (authored by pioneer Robert Boyle) were attempts to multiply eyewitnesses to the events that unfolded in the lab. In doing so, Boyle reified the prototype for research article generic constraints (e.g., a focus on methods and a careful discussion of results) that still exist in scientific reports in the field. "The reader was to be shown not just the products of experiments but their mode of construction and the contingencies affecting their performance, as if he were present" (Shapin 511). According to Shapin, Boyle cited and used visual representations and lab equipment accordingly. Visuals were "mimetic devices" (Shapin 492) for verifying that what he said actually happened, and pictures of instruments (Boyle's air pump, in particular) were rendered as a means to encourage the experiment's replication and, thereby, multiply eyewitnesses.

In *Communicating Science*, Gross, Harmon, and Reidy compared a sampling of twentieth century research articles to those gathered from the nineteenth, eighteenth, and seventeenth centuries. They found that the modern linguistic features developed over time and that older articles used fewer citations, fewer complex sentences (Gross, Harmon, and Reidy 167) and more figurative and other embellished language (Gross, Harmon, and Reidy 125 and 173). Methods sections didn't developed until the twentieth century (Gross, Harmon, and Reidy 184). Therefore, relatively speaking, methods sections (wherein authors explicitly describe their procedures) are a recent development. Gross, Harmon, and Reidy also made generalizations about typical contents of twentieth century articles with which to compare the SAS article. They gathered over six hundred ten-line passages (seventy-five of which were from astrophysics) from thirty-five scientific journals printed in the twentieth century and traced style and content issues common to the lot. They found general style and content characteristics shared by most of the samples (Gross, Harmon, and Reidy 161–213); for example, modern scientific articles tend to hedge claims with qualifiers such as "might" and "possibly." Modern research articles also cite other works often and tend to use passive voice to describe actions.

The methods section was of most interest to me because scientists usually use this section to discuss *how* and *why* they went about conducting their experiments. The methods section of the published SAS article does give some indication what rhetorical devices the team used to make their ad hoc decisions. However, most often, the article reports the decisions in a particular way—namely it uses passive voice and it contains sentences constructed in a way that the simulation itself is the subject of the decision made. Using passive voice transfers agency onto the object of an action, rather than the agent performing the action. Furthermore, writing a sentence where a phenomenon is the subject of the main clause attributes causal authority to the phenomenon, because subject-verb relationships often ascribe causation—the subject causes the object via the verb (Tarone et al.; Vande Kopple; Rundblad). Take, for example, this excerpt from the idealized methods section of the SAS article. This section discusses mathematical equations upon which TSI based their model. The team makes assumptions about the overall entropy (or tendency toward disorder) of the simulated supernova—my bold for emphasis:

> **The assumption** of a steady state isentropic flow **implies** that there is a zero gradient in the entropy of the postshock gas, which in turn implies **this flow is marginally stable** to convection. **This allows us** to separate effects due to convection from other aspects of the multidimensional fluid flow. (Blondin, Mezzacappa, and Demarino 972)

In this example, the team does not serve as the subject of any sentence. Instead, their assumption and the marginally stable flow imply and allow things to happen.

Further, the team describes setting initial values—my bold for emphasis:

This value of the boundary flow velocity **is chosen** based on the analytic value at the boundary radius (see Fig. 2), and, in effect, **this value sets the height** of the standing shock. (Blondin, Mezzacappa, and Demarino 973)

In this case, the team uses passive voice in such a way that it draws readers' attention to the values, densities, boundaries, and pressure gradients themselves, rather than the team's decisions to set those values. The sentences read as if the phenomena and variables predicated themselves. These ad hoc moves all fall under the categories of ad hoc decisions described by Winsberg—setting initial parameters, making creative decisions to add artificial elements, and deciding between symmetries to use. However, passive voice and use of phenomena as sentence subjects foreground the decisions made and background the decision process.

Prior drafts of the SAS article document many more of the team's rhetorical strategies—the earlier the draft, the more the team reported its deliberations. Many have written how published articles do not represent all aspects of scientific knowledge production and discovery (G. Myers; Latour). Randy Harris has discussed how scientific reports never tell every twist and turn of the goings-on in laboratories or out in the field. The SAS article is no exception; the TSI team considered others' work, other alternatives, and rhetorical figures to make ad hoc decisions. By revealing the earlier declarations, we get to see the rhetorical figures that the team used, beyond those in the formal report.

What, then, are the elements of the team's decision making? Upon what does the SAS team base their decisions when composing the simulation? The team enlists both physical evidence and rhetorical strategies. The next sections will describe some of the most important and prominent features of the drafts and code.

RHETORICAL STRATEGIES IMPACTING THE DRAFTS

The TSI made its most drastic changes between the first and second drafts and between the third and fourth drafts. An e-mail conversation between team members stimulated the prior set of changes; correspondence between the team and the journal reviewers accounted for the latter. The prior set of changes included major eliminations and additions; the latter, mostly additions. The introduction, numerical methods, 2-D results, and conclusions sections underwent the most alteration.

Solving Dilemmas

In the second draft, revisions, deletions, and additions serve to downplay the role that TSI played in making their own ad hoc decisions. In these two latter drafts, ad hoc changes are presented as unavoidable, irrefutable, or self-evident. First of all, in publication, the team strategically uses passive voice, and it makes phenomena, rather than themselves, the subject of sentences. For example, in the excerpt below, the team offers several explanations of ad hoc decisions in the numerical methods section from the first to the second draft. Notice how the team presents simulation parameters that they set as foregone conclusions with minimal explanation of why they use particular values to initialize the simulation and why certain values were held at fixed rates—my boldface for emphasis:

> **The** numerical **simulations are initialized** by mapping the analytic solution onto a grid of 300 radial zones extending from $0.1R_s$ to $2R_s$. **The radial width** of the zones **increases linearly** with radius such that the local zone shape is roughly square in the two-dimensional simulations, i.e., $\Delta R \approx R\Delta\theta$, to minimize truncation error. **The** two-dimensional **simulations used** 300 angular zones covering a polar angle from 0 to, assuming axisymmetry about the polar axis. (Blondin, Mezzacappa, and DeMarino 973).

This excerpt explains how the team initialized the numerical simulation. It also represents how the team used passive voice and phenomena as sentence subjects. In this case, the TSI team diminishes their role in initializing the simulation. By this account, the simulation itself—not team methods or decisions—justified setting these particular conditions.

In contrast, earlier drafts record more of the decision-making process that went into setting the ad hoc parameters. For example, earlier drafts contained a sentence that explained why the team chose the variables and parameters they did—namely, to offset a drawback of the algorithm that they used for the calculations. In one earlier draft, for example, the TSI team sets up a *dilemma* or double propositions wherein either they risk degrading either the integrity of their simulation or its similarity to familiar simulations (Lanham 54). In the next excerpt, the team offers explanations of why they made the computational decisions they made—my bold for emphasis on the explanation:

> The numerical **solution does depend slightly on the value** of the timestep . . . This dependence is relatively small, and **we are careful to construct grids and solutions such that the value . . . is similar in most simulations**.

Here, the team explains that the accuracy of the numerical equation they used might depend on values they set for how the simulation advances through time. They eliminate the *dilemma* from the next draft, and the passage reads as if no *dilemma* existed in the first place.

> If the initial conditions are set to match the inner boundary conditions . . . , our simulations are relatively uneventful. After initial transients die away, the solutions settle into a steady state that closely matches the analytic solution. (Blondin, Mezzacappa, and DeMarino 974)

Without the *dilemma*, the passage does not justify setting the initial conditions; it presents them as matters of fact or self-evident.

In the first draft, Blondin, Mezzacappa, and DeMarino set up another *dilemma* or double propositions wherein either they set a particular initial boundary, or they risk degrading their results. In the excerpt that follows, the team explains how they set boundaries that constrain their simulated supernova. Since a part of testing a supernova involves measuring how much energy escapes out of the system, they must set a leaky boundary out from which the energy can flow. The next excerpt explains why they made the leaky boundary as they did, despite the fact that it created possible "noise" or errors inherent to the boundary equations themselves. The team ultimately removes the *dilemma* from the numerical methods section; the earlier draft alleges that had the team not set an initial condition, more dire consequences would ensue—my bold for emphasis.

> *The pressure and density at this leaky boundary are adjusted to force zero gradients. . . .* ***(We could also have equated the pressure gradient and the gravitational acceleration, but the cancellation involved would most likely have proven too noisy for our purposes.)***

In this *dilemma*, the team chose one equation over another for setting values for the gas flowing out of the system. Had the team not set this initial condition, more dire consequences would ensue. Later drafts presented the decision to set the boundary affirmatively—as "influenced by the desire to construct time-dependent solutions" rather than as attempts to avoid complications such as noisiness.

Defending Choices with Diaelogia and Pareuresis

The team uses a parenthetical sentence to detail the weakness of an alternative method and, thereby, trumpet the virtue of the method that the team did decide to use. The parenthetical statement serves as justification as to why the team set the value the way that they did—it can be considered either a *dicaelogia* (offering a necessary or reasonable excuse) or *pareuresis* (a weighty excuse that overcomes objections) (Lanham 50 and 108). The team removes this sentence by the second draft:

> The pressure gradient at the inner boundary is set to match the local gravitational acceleration, mimicking a hydrostatic atmosphere (the SAS is nearly hydrostatic far below the shock). The density at this leaky boundary is adjusted

to force a zero gradient, allowing a variable mass flux off the inner edge of the numerical grid. (Blondin, Mezzacappa, and DeMarino 973)

In order to explain motivations behind setting these particular inner boundary conditions, the team keeps a sentence written in passive voice that distances them from their own methodological choice in the matter. Eliminated from the second and third drafts, these subjective implications no longer taint the passage. Now, the passage has a more objective tone. Whose desire influenced the setting of the boundary? Passive voice separates the TSI's team from its own decision making here.

Justifying with Compensatio

The team also eliminated other moments that exposed their deliberation process. Another passage in the second draft described why the team used a problematic algorithm. For example, they removed a sentence at the end of a paragraph that described a problematic algorithm in the numerical methods section. The point of the paragraph was to admit the drawback—namely, that it, too, caused noise in the system, noise that might have influenced or tainted their results. The eliminated sentence read as follows: "We find that this problem is worse in three dimensions than in two." The sentence was meant to offset the problematic algorithm vis-à-vis *compensatio*—weighing or counter-balancing one distasteful extreme with another (Lanham 38). Remember, the team only made 1- and 2-D simulations. When computer simulations reconstruct a complex system, they can do so in one-, two- or three-dimensions. Systems that look and behave the same all over translate into one dimension. Systems that have two identical sides require two-dimensions. Systems with heterogeneous, dissimilar parts require three dimensions (Winsberg, "Simulation," 31). The SAS simulation required two- and three-dimensional runs. With compensatio, even if the problem occurred in two dimensions, the problem wasn't nearly as bad in three dimensions, where TSI used it.

Explaining with Aetiologia and Concessio

The team also eliminates moments wherein they use first-person to justify ad hoc decisions. The first draft contained moments wherein the team justifies the simulation's initial conditions. For example, in a sentence included in the first draft (but absent from the second), the team explains that they must impose a leaky boundary as an initial condition for the simulation. "In this way, we ensure that perturbations to the setting solution, which would dominate this solution as we approached the origin, are advected off the grid." The team wanted to test if the standing accretion shock is what causes the star's energy to build up, so they had to eliminate any other factors that might affect the system's energy. They added the leaky boundary so that any other system disturbances would move off into the horizon (or advect) rather than feed or starve their simulated supernova. The sentence above explains why the team did what they did (by justifying and explaining the need for the leaky boundary); it is an example of a rhetorical move called *aetiologia*, wherein writers enumerate causes or reasons (Lanham 3). Nonetheless, they eliminate the explanation from the final drafts.

Further, they eliminated a paragraph justifying both symmetrical and creative ad hoc decisions. The first draft contained statements justifying some of the creative additions that the team made. In constructing the system's boundaries, they had to set constant accretion rate (e.g., the rate at which gasses and particles accumulate or build). The decision is a creative one, because it runs contrary to the accepted notion that core collapse supernova have decreasing accretion rates. In a parenthetical statement in the first draft—my bold for emphasis—Blondin, Mezzacappa, and DeMarino attempt to foresee and redress their audience's reservations about their decision:

> The fluid variables at the outer boundary are held fixed at values appropriate for highly supersonic free-fall at a constant mass accretion rate. **(While this does not reflect the decreasing accretion rate in a core collapse supernova, . . . a constant mass accretion rate is imposed to obtain an analytic solution. Nonetheless, this should not affect the conclusions drawn here.)**

Here, the team admits that they lose one virtue (a decreasing accretion rate similar to core collapse supernova) in order to gain another (an analytical solution). They also assure the reader that this difference (between the supernova's actual accretion rate and the one that they model here) will not affect their conclusions about whether accretion shocks do, indeed, cause an instability that helps stars explode. The move typifies *concessio*, or conceding a point to prepare for a more important one (Lanham 38). By eliminating the *concessio*, the team distracts the reader from weighing the alternatives.

Narrating with Pragmatographia and Progressio

The second and third drafts are also where the TSI team makes the bulk of additions to the conclusion. The original draft simply reports the presence of instability in the standing accretion shock. It spends only one citation-filled paragraph describing why; such thorough explanations are called *pragmatographia*—vivid description of actions and events (Lanham 118).

> [W]e note that the amplitude and scale of our seed perturbations are consistent with those expected to be produced by the late stages of shell burning in pre-supernova stars (Bazan and Arnett 1998) or in the core-collapse process itself. . . . Alternatively, entropy-driven postshock convection (Herant et al.

1994, Burrows, Hayes, and Fryxell 1995, Janka and Mueller 1996, Mezzacappa et al. 1998, Fryer and Heger 2000) could provide the necessary seeds for this shock instability. Yet another possibility is the growth of lepton-driven convection (Keil, Janka, and Mueller 1996) or neutron fingers (Wilson and Mayle 1993) deep in the proto-neutron star. (Blondin, Mezzacappa, and De-Marino 979)

There was an increase in citations in sections describing ad hoc decisions. Here, we see why. The team increasingly used others' work to account for their own decisions. Rather than draw their own conclusions, the team summarizes others; or rather, it draws conclusions by applying and comparing its results and implications to others, a rhetorical move called *progressio* or building a point around comparisons to other works (Lanham 129). In later drafts, they also draw *analogies* between their work and others' to contrast expected from observed outcomes: "Herant, Benz, & Colgate (1992) pointed out that the scale of convection in a confined spherical shell scales. . . . By analogy we might then expect the dominant mode of instability in the accretion shock to scale with R" (Blondin, Mezzacappa, and DeMarino 977). The team compares its work not only to others in the field, but also their own prior publications. By the second draft, the conclusion includes the following new paragraph about an article the team published years earlier.

> In Mezzacappa et al. (1998), once the shock front is distorted by the initial neutrino-driven convection, the SAS instability does appear to affect the post-shock flow a few hundred milliseconds after bounce. . . . However, unlike in our SAS models, the 1 1/4 1 mode in fact did not come to dominate the postshock flow. (Blondin, Mezzacappa, and DeMarino 979)

Here, the team speaks to their own previous research in order to contextualize the current study and frame the alternative paradigm. It appears that the simulation scientists draw conclusions during the drafting process (between the first and second drafts). In chapter 5, we'll see how the "bottom line" emerges from e-mail conversations. The team's new findings also help them reinterpret older studies. They cite other scientists' related work to reinterpret older work in light of their new findings. The SAS findings help explain the phenomena that the others' identified.

> It has been pointed out that the energies of explosions powered by neutrino heating may be limited to $(1–3) \times 10^{51}$ ergs (Janka 2001). If the SAS instability sets in and also serves to power the supernova, we may wind up with a more energetic explosion, one only partially powered by neutrinos. (Blondin, Mezzacappa, and DeMarino 980)

How the authors use others' research here falls in line with prior work in the rhetoric of science about the rhetorical nature of physical evidence and scientific facts. Latour and Woolgar discuss how scientific facts have meaning through socially constructed networks or sets of positions in a discipline. Also, Danette Paul describes how scientists cite facts from prior research for specific rhetorical ends—for reinforcing their level of certainty and for creating a research space upon which their audience can agree and from which they can judge their claims. In this case, the SAS team uses well-established facts from physics to scaffold their new claims about an alternative mechanism for supernovae explosions. They cite widely accepted concepts and citations to explain and justify the new mechanism. Overall, the later drafts contained more citations than earlier drafts—in the introduction and conclusions in particular. With each draft, the team found more and more ways to frame its work in the context of others' work.

Foreseeing Questions with Hypophora and Negatio

Further, while all drafts of the conclusion contain passages that ask multiple lingering questions ripe for others to investigate, the team adds possible answers to the third and fourth drafts:

> How will rotation and/or magnetic fields couple to this mechanism? Will rotation provide sufficient distortion of the accretion shock. . . ? And will this lead to an explosion where otherwise (in the absence of rotation) one would not occur? It may be that a combination of near-explosive conditions and rotation will be required before the instability we describe can develop. (Blondin, Mezzacappa, and DeMarino 980)

Closing open-ended questions by listing possible answers is called *hypophora* (Lanham 87). The team achieves a similar outcome when it adds *negatio* or denial (Lanham 103) to its conclusion to describe what its results don't show: "It is important to note that this behavior is not driven by convection. . . . [T]here is no global gradient extending over the entire postshock region, driving convection on that scale" (Blondin, Mezzacappa, and DeMarino 978). Both *hypophora* and *negatio* direct readers' attentions to particular claims—the former does so by announcing the most favorable of all possibilities; the latter does so by eliminating unfavorable alternatives.

Whereas the second and third drafts eliminate alternatives, present self-evident justifications, and draw out far-reaching and groundbreaking conclusions, most of the changes to the fourth draft—the one that precedes the published article—increase hedges and citations, expand explanations, and make other changes to qualify TSI's work. For example, the team adds two new citations to the introduction in this draft in order to frame the groundbreaking news.

[A]n analysis of a converging blast wave suggests that such shocks are unstable for high Mach numbers (e.g., Whitham 1974), in contrast to planar shocks. More recently, Foglizzo (2002) has described a vortical-acoustic instability of spherical accretion shocks in the context of accreting black holes. (Blondin, Mezzacappa, and DeMarino 971)

The team foreshadows their big discovery—namely, that instabilities in the accretion shock might fuel supernova explosions—by pointing to "feedback" discovered in someone else's work. The team also emphasizes its findings by contrasting its work with others. One peer study analyzed what happens when accretion shocks decrease in temperature rather than vorticity. The TSI team ran the peer study's configurations through its own simulation. In certain cases, the peer study bore out TSI's results. "However, for $1/4 \ 4=3$, Houck & Chevalier (1992) did not extend their analysis to the nonradial case; therefore, we cannot directly compare their results with our findings in this case" (Blondin, Mezzacappa, and DeMarino 979). In this instance, the team uses citation to distance itself from another study in the field; in doing so, it establishes their work's uniqueness and validity by default.

Using Abduction

TSI's conclusions develop via abduction. Remember from a prior chapter that Peirce describes abduction as an alternative to induction and deduction in science. Induction, deduction, and abduction would treat the same premises differently. Take, for example, the different conclusions you can draw from different combinations of similar facts. Deduction takes what is true of the general case and applies it to a specific case: All humans die (general, major premise). Socrates died (minor premise). Therefore he was human (specific conclusion). Induction generalizes facts from the specific case: Socrates was human (specific, major premise) and he died (minor premise). So, all humans must die (general conclusion). Abduction interchanges the conclusion for what would be considered a premise of an inductive or deductive syllogism: *All humans die (general, major claim); Socrates was human (specific, minor claim), so he died (conclusion).* In deduction and induction, the strength of one or the other premise warrants making the leap to the conclusion. If the general premise is true, then it strengthens the specific conclusion. Abduction yields problematic warrants—warrants that, from a symbolic logical perspective, are not necessarily as strong and premises that are not tautologies. Abductive statements are not necessarily true; they don't have the logical form that resists falsehood. For example, the team uses abduction to brainstorm a major conclusion. Emails between the authors included the following:

If the explosion is indeed "jet-like" . . . , what would the ejecta density distribution look like at late times—i.e., hundreds of years later as in Cas A?

Here, a general premise (explosions are "jet-like") and specific premise (what ejecta might look like in cases like Cas A) are linked together. However, unlike in induction or deduction, in the case of abduction, neither the general nor specific premises have been proven beyond the shadow of a doubt. And, these contingent premises are used to make hypothetical conclusions.

Overall

The published draft revealed the team presenting a confident ethos, air-tight reasoning, and no other alternatives but the ones they chose. The working drafts tell another story, one in which the team's own decisions and choices impacted their work. The team gravitated away from rhetorical moves that presented alternative ad hoc scenarios and toward those that depict their ad hoc decisions as foregone conclusions. (See table 4.2.)

The earlier drafts reveal some of the rhetorical strategies—*deliberatio, dicaelogia, compensatio, concession*, and others—that went into making the ad hoc decisions. The later drafts, on the other hand, eliminated much of that deliberation and replaced it with rhetorical moves—passive voice and *aetiologia*, for example—that somewhat obscured the decision-making process. Also, latter drafts contain moves—*hypophora* and *negatio*, for example— that answer open-ended questions and left less room for questions in readers' minds. Later, I'll discuss how the genre of the scientific article may have suppressed the team's ability to report all of its rhetorical strategies underpinning the decision making that went into creating the simulation. Examining prior drafts reveals several of the rhetorical strategies that grounded ad hoc programming.

Table 4.2. Rhetorical Moves Dominant in Drafts

Draft 1	Draft 2	Draft 4	Published
• deliberatio	• pragmatographia	• citations	• pragmatographia
• concessio	• negatio	• passive aetiologia	• passive aetiologia
• pareuresis	• hypophora		
• dicaelogia	• progressio		
	• analogy		

NARRATING THE CODE

Earlier, we saw how the team began building a story about their simulation. They gradually eliminated narrative options and argumentative options, until the SAS article reported a cohesive story about ad hoc decisions predicated by phenomena rather than team decisions. Rhetorical figures in SAS article drafts are not the only means by which the team crafted the story of the SAS simulation. Novel features of the simulation code itself also aided in the storytelling. Blondin used FORTRAN to develop Virginia Hydrodynamics #1 (VH1), the software an adaptation of which the TSI team used for the simulated supernova in question. Almost a third of scientific researchers and design engineers use FORTRAN because of its convenience for numerical computing due to its similarity with algebra (Kupferschmid 9). I compared two different versions of the code—one posted online in 1991 and the other in 1998. In particular, I analyzed the differences between how the two versions handled comments within the code and semantic cues.

Comments are natural-language guideposts that programmers write within code to explain a sub-routine's function; computer scientists have yet to reach consensus on when and how to use them. Lawrence Weissman found that comments didn't aid in comprehension. Kernighan and Plauger warn that useless and intricate comments stem from sloppy code. "[W]hen comments become too involved, ask whether the code itself is at fault. . . . Don't comment bad code—rewrite it" (Kernighan and Plauger 143–144). Others disagree; they find great value and purpose in using commenting. Diomidis Spinellis disagrees that comments are crutches for sloppy programmers. He tells an anecdote about well-commented complex IBM software that taught him volumes about good programming style. He describes how the "first large production-grade software" had "6,000 lines of 8086 assembly code," which "could have easily been a real mess" were it not for the fact that almost "every line was commented," which "ensured that a program written in a primitive language was actually readable" (Spinellis 87). Spinellis—who recommends using comments per every ten lines of code—argues that comments aid code comprehension as much as indentation and other formal features do. He warns against poorly written comments that do no more than recap the code that they're describing (Spinellis 87), and he recommends postponing writing comments until after writing the code (Spinellis 89). Elshoff also warns that external documentation about code tends to be outdated; therefore, comments serve as the most immediate and accurate instructions about the code, particularly software updated on a regular basis. "This difficulty [updating code] can be reduced for all future modifications by adding appropriate comments as discoveries about the program are made" (Elshoff 513).

A few studies and stylebooks recommend and analyze typical comment format. Usually comments are written in imperative or indicative mode to indicate to programmers the purpose of the sub-routine (Hughes-Etzkorn). They also serve different functions; some define sub-routine content or elements (definitional), some describe what the sub-routine produces (descriptive), some give programmers instructions about sub-routines to add or other maintenance to perform (instructional), and others explain how a sub-routine operates (operational) (Hughes-Etzkorn 79). A popular programming stylebook recommends that programmers use comments to "help the reader over difficult spots in a program" (Kernighan and Plauger 143).

Regarding VH1, the old version of code contained 360 comments in 1554 lines—about two comments per ten lines of code. The new VH1 contained 535 comments in 2,197 lines of code—again approximately two comments per code line. However, the new version included comments featured in entirely new sub-routines. Of shared sub-routines that exist in both old and new versions, there was a decrease in comments. Revised sub-routines contained 312 comments within 1,307 lines. Numbers of lines of code also decreased in sub-routines common to both versions of VH1. Although there was a decrease in total clauses (360 to 312) from the old to the new common sub-routines, there was an increase in operational (from 34 to 50 comments), definition (from 174 to 307), and descriptive (from 100 to 146). Instructions, surprisingly, decreased in number (from 48 to 42 comments). Also, most of the comments in both old and new versions were written in imperative mood (31 percent in old, 60 percent in revised, and 47 percent in new overall). But there was an increase in declarative mood from old to new versions.

Although most comments usually define or detail the function of the subroutines that they precede, comments in VH1 had unorthodox traits. For example, the comments contained in-text citations to others' work. I found that both old and new versions shared 3 sub-routines—in parabola.f, riemann.f, paraset.f—with comments that included APA citations to journal articles that inspired the underlying architecture. Here's one from the initial comment in parabola.f: "Colella and Woodward, JCompPhys 54, 174-201 (1984) eq 1.5-1.8,1.10 // parabola calculates the parabolas themselves. Call paraset first for a given grid-spacing to set the constants, which can be reused each time parabola is called." In this case, the team frames operational and descriptive comments with a citation from the *Journal of Computational Physics*. The citation also serves to show the reader that the programmers had backing from two others in the scientific community. Here, the VH1 program also invites users to see for themselves how the original programmers derived the routine. Other programmers who use the code for their own purposes already know that the programmers used prior knowledge in composing the sub-routine. Now, they also can identify the very article or book upon which the sub-routine is based. Including citations implies that the code is a product of research. Likewise, in riemann.f, the team cites the article that inspired how they applied a well-known equation: "Solve the Riemann shock tube problem for the left and right input states, C using the Newton iteration procedure described in van Leer (1979)." Here, the team not only invites users to go back to the source of the sub-routine, they also write a descriptive command in imperative mood, thereby linking the sub-routine's functions with researchers' efforts and users' deliberations. Citations in the comments also go to show that scientists often borrow ad hoc decisions and other features of other simulations. Other simulations stand as a basis for justifying using particular algorithms and imposing initial parameters.

Furthermore, both old and new initial comments in shared sub-routines contained creative (and seemingly critical) word choice or terminology. For example, the team imposes what it calls "fictitious forces" (e.g., creative additions) in evolve.f and forces.f. Initial conditions were called "phony boundaries" in flatten.f. In init.f—a sub-routine meant to check others—a certain check was parenthetically called "a whimpy test." And in init_sss, the programmers admit that one of the sections of the sub-routine is the "the Achilles heel of PPM [which stands for piecewise parabolic method]." (PPM is an algorithm commonly used to construct two-dimensional grids upon which the TSI team runs its simulations.) Later on, in the published SAS article where they describe this particular computational method, the team admits that "a drawback of this algorithm is the generation of postshock noise when a strong shock aligned with the numerical grid moves very slowly across the grid. This is exactly the situation encountered in this problem" (Blondin, Mezzacappa, and DeMarino 973). By using figurative language and calling the PPM "the Achilles heel of PPM," not only does the team communicate what the sub-routine does, they also forcefully convey what it doesn't do and how it has the potential to weaken the entire program. By acknowledging weakness, they show thoughtfulness and strengthen their case.

Citations and figurative language are not the only unique features I found. Narrative also factors prominently in the VH1 comments. If you recall, narrative was not among the discursive modes that Hughes-Etzkorn said typify comments (e.g., description, definition, operational, instructional). Almost all of the new sub-routines that set initial conditions (init_bow, init_hzs, init_khi, init_sod, init_sss, init_stb, init_swb and init_f) describe the phenomena replicated by the code, but with a difference. Hughes-Etzkorn describes how a typical comment might read as follows: "This routine reads data." However, VH1, one of TSI's modeling software, includes multiple narrative comments: "A supersonic wind originates at the surface, driving a bubble into the surrounding medium. 2-D spherical (R-theta)." This comment (and the others just like it) describes the action of the supernova rather than the sub-routine. Taken together, they qualify as a narrative about the phenomena that the sub-routines are meant to represent.

The TSI team uses metaphorical language and citations, thereby anticipating and inviting future users to exercise their agency and prerogative to check and change the sub-routines. These unconventional features of the comments help make the simulation code more transparent than traditional code, which often only explains what sub-routines do, not how they do it. The audience of the VH1 software—the users who'll make their own simulations with the software—are invited to compare the code to the sources. They are invited to visualize what the sub-routines create (rather than just how they churn numbers) and, thereby, imagine alternative methods and system behaviors.

DISCUSSION

Rhetorical figures underpin the ad hoc decisions made by computer simulation scientists. The TSI team made concessions, identified dilemmas, used narrative and other strategies to bring the supernova mechanism before our eyes. TSI weighed alternative findings and options in creating the supernova simulation. Rhetorical tactics and argumentation are necessary in creating the virtual representation of actual phenomena. Recognizing and studying the rhetorical underpinnings of simulation production also can help make inroads into formulating a theory of reception that specifies how simulations are interpreted, appropriated, used, misused, proven and disproven. It's clear that the simulation scientists are doing more than simple math. They're building a story about phenomena, and they use what is at their disposal—including comments—to build the most convincing story.

The published manuscript did not include many of the rhetorical figures used to justify the ad hoc decisions, so readers did not have as much access to these important insights into the simulation design. However, published scientific articles rarely report all deliberations. Research shows that published articles do not represent all aspects of scientific knowledge production and discovery. Scientific articles rarely give the fullest accounting of what happens in the lab. Indeed, the linear narrative that scientists construct in articles excludes many of the twists and turns—i.e., missteps, tangential details, fruitless deliberations and other occurrences—that occur in the lab and directly and indirectly help scientists make decisions and draw conclusions. Likewise, rhetorical decisions help simulation scientists choose the best path in decision making about what to include and exclude from their simulation and calculations.

As is typical of science writing, the drafts evolved to background the authors' deliberations and foreground the simulated phenomena. In time, they gradually emphasized how the simulation itself necessitated setting certain parameters and initial conditions. However, the earlier rhetorical figures depicted the team deliberating between options. Many of the justifications for using one option over another happened by process of eliminating more subjective or intuitive or less objective or direct alternatives. Even when observational data are incomplete, sound and reasonable logic must underpin ad hoc decision making. In the SAS case, their simulation of an actual supernova might seem arbitrary because they cannot compare it to the real thing, but it is not arbitrary or implausible because good reasons support their decision making. It may be advisable for journals to change author's guidelines and editorial standards for scientific articles that report simulations. By allowing for longer word and page limits and recommending more explanation when reporting the rhetorical choices behind the simulation, journal editors might foster better understanding among colleagues in the scientific community and help prevent misunderstandings like Climategate from escalating.

Rhetorical choice is so integral to simulation programming that a failure of rhetoric can cast doubt on the findings that the simulation produces. Concerns about climate change simulations are a part of what fuelled the "Climategate" controversy in November 2009, when e-mails and documents from the University of East Anglia's Climatic Research Unit (CRU)—one of the world's leading research facilities that publishes data proving global warming—were leaked on the Internet. The documents that climate skeptics found most troubling were e-mails wherein CRU researchers call one of their algorithms a "trick" used to "hide the decline" in modern temperature amidst an overall temperature increase (Pearce). Climate skeptics heralded this language as proof of misdeed on the part of the CRU team. However, independent investigations found that those seemingly deceitful terms were "colloquialisms" for orthodox statistical and computer programming methods commonly used to reconcile disparate data sets (Randerson et al.). The team "hid" the results only insofar as the internal logic of the simulation required. Chapter 6 will discuss more about controversies surrounding climate change simulations. The TSI case shows that these kinds of deliberations are not disingenuous or purely semantic, but rhetorical and necessary for computer simulation methodology and integrity.

In some ways, the SAS project does not align with some standards expected of scientific texts. Gragson and Selzer describe how most scientific texts contain certain linguistic moves "in keeping with the norms of the objective scientists . . . nominalizations . . . explanations [couched] in the passive, and . . . awkwardly impersonal constructions" (31). On the contrary, the SAS article does not use as many citations, hedges, or passive constructions. It interjects questions and exclamations in the conclusion (see chapter 5) to dramatize the news about standing accretion shocks in supernova. Still, the paper also does somewhat conform to standards, insofar as many ad hoc choices are underreported. Gragson and Selzer also researched the motive behind using passive constructions: collegiality. They describe how Maynard Smith defends a longstanding, but controversial theory; when Smith uses passive voice "he is not so much reminding his reader of information they . . . might have forgotten as he is asking them to place themselves within the community of shared beliefs" (Gragson and Selzer 31). If passive voice works the same way in the SAS article, then the TSI team is assuming that their decisions will come as no surprise to whoever reads the paper. The team is also assuming that whoever reads the paper will be less interested with their actions (and how they tinker with the simulation) than with using the simulation findings for furthering their own endeavors. If, as Gragson and Selzer imply, word choice, conformity to generic standard and other rhetorical and linguistic cues are indication of authorial intent to forge new ground or validate old ground, then the SAS article demonstrates how TSI intends to materialize the phenomena or simulation. They use the linguistic features—active verbs, fewer passive moments—that substantiate the phenomena and the simulation in the minds of readers.

This case somewhat counters Thomas Kuhn's ideas about sudden and volatile paradigm shifts. Much the way that John Angus Campbell wrote about the religious underpinnings in Charles Darwin's work, this case of reporting in computational science used mostly conventional writing to share virtual evidence and propose revolutionary ideas. According to Campbell, Darwin used an analogy between natural selection and domestic animal breeding (209). He also described evolution as means by which "current unpleasant natural states" progress "along the way to more advanced life" that the Creator intended (213). He claimed that time would expunge abhorrent behaviors (cannibalism, for example) observed in certain species. Even though the former analogy oversimplified natural selection almost to the point of error, it helped Darwin persuade his agrarian contemporaries with a seemingly common sense appeal. Further, references to a "more advanced life" that God intended helped sell evolution to those with religious inclinations. Similarly, if the SAS article altered astrophysical science, then it did so with honey rather than vinegar or vitriol. The earlier rhetorical figures depicted the team deliberating between options not otherwise equal; many of the justifications for using one option over another happened by process of eliminating weaker, less attractive alternatives.

The SAS case demonstrates how the rhetorical practices and genres of the scientific community play a significant part in effacing or erasing any traces of rhetorical production, a production trail that exposes the extent to which simulations do not and cannot perfectly recreate actual events. That the SAS article did not disclose more about the deliberation that went into it (the deliberation that clearly makes the simulation a rhetorical artifact) was not the fault of the computer simulation itself so much as what the scientific community has come to expect (and make) of the genre of journal articles, the process of manuscript review, and the practice of citation and reference. The SAS article evolved to contain fewer rhetorical figures most likely due to the strictures of the scientific ethos—its compulsion to present itself as objective rather than subjective, definitive rather than experimental.

The case of the SAS simulation illustrates some of the rhetorical mechanisms responsible for making some simulations better than others. It appears that a simulation must not only be based on sound data, but also sound reasoning, argumentation and a convincing story. Newer theories demonstrate how computer technology manifests ideological and hegemonic tendencies. Galloway, for example, demonstrated the hegemonic nature of the protocols and equations that govern exchanges in information technology. Protocols impose a tyranny of their own insofar as they impose rules and constraints that delimit how we can transfer data and to what networks we are privy. Similarly, the equations and values that govern simulations reflect deliberate decisions made from weighing options, information from past research, choices about which options are the lesser of two evils, and so on. And generic constraints and preestablished writing practices within the scientific discursive community delimit how news of the simulation is reported, received, and recycled.

Eric Winsberg describes the difficulties that simulation scientists currently face determining the validity and verifiability of their simulations. When it comes to simulations, verification (e.g., checking that simulations correctly represent the target system) and validation (e.g., checking that the model produces accurate numbers) are more closely intertwined than for small-scale models of simple systems. "What simulation scientists are forced to do is to focus . . . on establishing the combined effect of the models they begin with, and the computational methods they employ, [and] provide results that are reliable enough for the purposes to which they intend to put them" (Winsberg *Computer Simulation* 838). For example, global climate is too complex to produce a computer model that perfectly matches the theories about climate change and the actual climate itself. "We will have to try and show, instead, that the combined effect of our choice of model and the computational methods produce predictions that are good enough for our policy-making needs" (Winsberg *Computer Simulation*, 838) But by what measure should we determine which predictions are "good" and "reliable" enough? Rhetoric affords us the tools necessary for negotiating and evaluating premises and arguments in politics, business, advertising, and other realms; I contend that it can do so in ad hoc modeling, as well. It is rhetorical strategy that helps persuade an audience that our choice of methods, decision to include or exclude concepts, and other deliberations are sound.

The field of computational science could benefit from more rhetorical analyses of its methods. Rhetorical education and a more transparent reporting of rhetorical methods employed could serve to illuminate and improve the science of creating computer simulation. Unfortunately, many introductory and advanced sections of simulation courses do not contain units wherein the class discusses ad hoc or rhetorical skills necessary for building simulations (Fell, Proulx, and Casey; Shiflet). Some have found that rhetoric improves how instructors teach programming and how they write active-learning-oriented assignments involving interactions with clients (Wolfe). I contend that the necessity of rhetorical know-how does not end for college-level computer programmers, nor does it solely train them for dealing with actual clients. It does not merely make assignments more contextualized, interesting, and engaging. Rhetorical strategies help rationalize the ad hoc reasoning necessary for creating simulation programs, and they help improve and expose more of the methodology and deliberation that produces the simulation.

In summary, the TSI team made ad hoc decisions in terms of rhetorical figures. Even when observational data are incomplete, sound, reasonable assumptions must underpin ad hoc decision making. The published SAS draft used rhetorical figures to account for the ad hoc decisions, even though it emphasized self-evident rhetorical figures and downplayed deliberative ones. In this way, rhetoric factors prominently in the making of simulation knowledge.

Chapter Five

Social Dimensions of Simulation Meaning

In the prior chapter, we saw how the SAS project entailed several rhetorical moves underpinning their ad hoc decisions. We also saw how drafts of the research article and the code itself both collaborated to tell a convincing story about the simulation phenomena (over their methods). This chapter will explore how the SAS project also means social interactions that bear upon the production and distribution of the project. First, we see how counterargument and abduction play a part in how the team draws conclusions. Next, we explore traces of how organizations, the scientific community, and media coverage transform virtual evidence into information for practical application. The team reframes the SAS project in different contexts—presentations, newsletters, annual reports, and other documents. Ultimately, what resonates most with the scientific community is the virtue of the simulated supernova itself, because the community uses it to ground decisions for other models and simulations. And the general public finds the most practical applications or potential for public good most interesting from the SAS findings. The documents surrounding the SAS project expose the role of decision making and social realities in the reporting of virtual evidence. They also point out the relationship between practical accountability and virtual findings.

THE TEAM NEGOTIATES AND COLLABORATES

Interpersonal relationships and communication between team members affected the scope of the SAS article. E-mail conversations between team members offer evidence about how counterargument and abduction factored prominently in deciding content to include in the SAS report. I use e-mail correspondences in a way similar to talk-aloud protocols. As in talk-aloud protocols, writers often reveal their intentions and motivations in e-mails. Jone Rymer used talk-aloud protocols to probe scientists' decision-making processes through multiple drafts of articles. Silvia Bernardini summarizes the work of Ericsson and Simon, theorists who identified essential features of talk aloud protocols telling of cognitive processes. The method works best when (1) the subject announces immediate impressions stored in his or her short-term memory and (2) researchers keep interaction with subjects to a minimum (Bernardini 2). E-mail correspondence during drafting shares the same merits. First, the TSI team composed the e-mails *in media res* while writing the drafts, so the e-mails often catch their ponderings, excitement, and doubts about claims. The TSI team e-mailed one another *while* they drafted; many of the messages were written after reading each draft. The e-mails also have moments when the team volleyed messages within hours and minutes of one another. Second, the e-mails were recorded without my observation, so no researcher-subject interaction took place. Much like talk aloud protocols, e-mail records stand as a present, written account of past impromptu conversation.

The first round of TSI e-mail transmissions shows the team working collaboratively and rhetorically. The team corresponded from the end of February to the beginning of March 2002; in these e-mails, Blondin and Mezzacappa correspond about the original draft (absent from the e-mails). In the second round of e-mail transmissions from June 2002 to July 2002, we see how conclusions about their work develop *as* they draft, rather than direct *how* they draft; the team discusses major revisions in these two months.

The team began drafting to accommodate audience concerns well before entering the peer review process or publishing the SAS article. For example, the team adjusted the scope of the paper to make the research sound more narrow, applicable, and thereby, appealing. In the article, the team justified its research as a continuation of studies on supernova explosion mechanisms. The team wanted to show that they added a new possible explanation for what fuels supernova explosions, but they wanted to do so in such a way that made the finding universally appealing. So, in e-mails, Mezzacappa asks about the scope of the article: "If you want it to remain general, then we probably shouldn't add too much that is specific to a particular application, unless we add an equivalent amount for all applications." Mezzacappa knows that scope determines content specifics. Also, Mezzacappa asks Blondin to reconsider the title and abstract of the paper in such a way that inscribes the most useful scope and purpose.

> The title and abstract and overall gestalt of the paper are quite general. . . . Does this work have its greatest impact in core collapse supernova theory? If so, it might be better to tailor the title, etc., to the supernova problem.

Mezzacappa reminds Blondin of the rhetorical importance of titles and abstracts; he suggests they rename the paper and rework the abstract so as to make the greatest impact in the field. Linguist John Swales found that most researchers use introductions to assess the territory of a particular field and poise themselves as unique in the field of scientific research. Here, the TSI team is aware that titles and abstracts also serve that purpose. This evidence from e-mail conversations shows that before and into the drafting process, the SAS story continued to take shape.

Mezzacappa also displays a keen sense of audience very early on in the team's e-mail conversations. He asks Blondin to expand the description of methods and give more detail explaining why certain decisions were made by adding sentences that rationalize a value they used for a certain parameter. Mezzacappa also wanted the draft to answer the question, "Where does the radial dependence in the second term of equation five come from?" Addressing a particular value set, he asks, "Is there a way to see/explain why there is a critical value of gamma (1.522) below and above which the behavior of the solution is different? It might be good to say something brief and intuitive." Mezzacappa's line of questioning resembles counterargument techniques; he has audience members in mind who have enough astrophysical knowledge to accept intuitive explanations and require them often. Saying "something intuitive" would enlist his audience's common sense. In summary, Mezzacappa plays devil's advocate, predicting and questioning areas in the original draft where audiences might have concerns.

Specifically speaking, Mezzacappa uses *praemunitio* or strengthening beforehand and defending in anticipation of attack (Lanham 116, 118, 120). He finds unclear moments begging for further elaboration. For example, Mezzacappa asks Blondin to explain his handling of initial versus boundary conditions: "When you say 'matched,' do you mean that the analytical solution on the grid is set so that the values at the boundary are equal to the values you specify . . . in your boundary conditions?" If the sentence was unclear to Mezzacappa, then an outside audience might have similar questions. Remember, Mezzacappa was a part of the SAS team. He understood from experience the decisions that went into the methods. The reason why he asked the question of Blondin was not because he didn't know or remember why they used those methods. He asked the questions because he had mind that they were explaining their methodology to outside audiences. He also uses *prolepsis*; he anticipates objections (Lanham 120). For example, he requests a fuller account of results that he knows will raise a red flag in the community: "In Section 4, we definitely need to somehow document what happened when the shock was perturbed with other values of 1 (you only mention 1 = 1)." Mezzacappa predicted that audiences will need the specifics, so he wanted the paper to explain further. He made the suggestion not because he didn't already know the answer—he's a part of the team. He felt audiences would question what happened at particular points in their argument, and he urged for more explanation now, rather than face criticism or misunderstanding later.

Working through drafts and rereading both the text and the visuals also gave Mezzacappa a new perspective on the significance of the team's finding. In turn, he requested content changes. In one instance, Mezzacappa reflects on figures that alter his thinking about their results (my bold for emphasis).

> **Figure caption 5 got me to think about something**. . . . If any negative gradients develop, the regions could become convectively unstable. We have no discussion of this point in the paper. Also, have you plotted the angle-averaged entropy versus radius in the 2-D case? **This might be instructive**.

He mulls over the figure and, while reading and interpreting it, he thinks of another point to develop in the paper. He suggests accounting for the possibility that entropy gradients feed the instability. The next draft includes just such a paragraph. In sum, the team develops new points and claims about their work as they review it and write about it. In this case, the prospect of audience counterarguments does not inspire new claims, but rather the process of reading and reflecting on a figure does. Reviewing the SAS article also inspired new perspectives on the team's prior publications. Upon reflection, Blondin offers up some possible implications of their work.

> Looking back at my paper on axisymmetric supernovae (ApJ, 472, 257), we almost totally neglected the case of asymmetric ejecta. In particular, we only tried the case of higher density in the equator rather than the pole, and we didn't mention anything about the geometry of the reverse shock. There may be room for some new work here.

Blondin remembers that his prior work mentioned nothing about asymmetric ejecta, and he ponders the implications of their findings on current understanding of observational data on Cas A (a supernova). The SAS article discusses those ejecta and reveals them as a component of accretion shocks. New findings cast older findings in a new light and reveal a lingering question worth investigation in future research. Reviewing his new work (the SAS article) uncovered a conspicuous gap in old work. The gap, in turn, inspires new implications for the SAS article. Past work provides a recursive springboard from which new ideas are launched.

Furthermore, the e-mail deliberations directly produce a distinctive detail of the published draft. The most exciting news about the simulation—that "[t]his instability exists!" (Blondin, Mezzacappa, and DeMarino 17)—emerges from team e-mail chats rather than from the final draft or in the computer lab. Their conclusions develop not only recursively (as the team read drafts), but also with fervor and via abduction. Upon reviewing the second draft, it comes to Mezzacappa that the team is onto something new. As he read and reflected on the draft, he began thinking about exciting new conclusions to draw from them. The possibilities so excited Mezzacappa that they kept him up at night: "I was up to about 1:30, in part because I was getting excited about some of the implications of our SAS model. I'd like to run those by you today, by phone, and I can put some of the discussion into the paper if you agree with the interpretations." Here, emotion factors into the creative process. Whether or not Mezzacappa's exhilaration bore the idea or vice versa, feelings heightened the idea's potential and provoked him to invest long hours into the idea. These long hours, in turn, inspired changes from the first to the second draft of the SAS article conclusion. Reading an early draft of the paper inspired Mezzacappa to begin thinking differently about their work. Mezzacappa not only suggests possible conclusions—the team doubled the size of the conclusion section to include it—he also requests additional simulation runs under various new conditions to help confirm the conclusion. In order to verify or substantiate these hunches, Mezzacappa requests a linear stability analysis, which is another round of simulation runs. With or without the runs, however, he suspects the findings significant enough to include in the *Astrophysical Journal*.

Abduction plays a part in helping the team uncover the most important finding of their work—namely, that the standing accretion shocks exist and play a part in supernova explosions. Mezzacappa continues in another e-mail on the same day, meant to go over the "big items," such as how to frame their conclusions. In this e-mail, Mezzacappa broaches several issues that set the course for the next two drafts of the SAS article (my bold for emphasis). He attempts to make sense of old findings vis-à-vis their new ones:

> What exactly is our instability? If an SAS is unstable, shouldn't it be unstable to ANY perturbation and under any circumstances? This is what we see in our models. **In the realistic case (e.g., in Mezzacappa et al. 1998) do we not have a "true" SAS as defined and explored in this paper and, therefore, we see "the instability" only under the right conditions? If so, this is part of the key to understanding what happened in Mezzacappa et al. (1998). . . . [W]hat happened in Mezzacappa et al. (1998)?** Did the neutrino cooling kill us by damping the waves, or did the postshock volume evolve so rapidly so as to increase the value of l so quickly that nothing ever developed, or both? Or did we never have a true SAS and, therefore, should never have expected an instability to develop and kick in (along the lines of (1) above)? . . . **Do the neutrinos significantly power the explosion as well, or is the redistribution of energy caused by the instability the key. Certainly they both contribute. . . .** Certainly the neutrinos are important, for many reasons, as we have discussed, but how much of the final explosion energy is due to them? **Before, the neutrinos were the vehicle for transferring the binding energy to the shock. Now, we have a different vehicle!**

In this e-mail message, the direction of the SAS article moves toward the team's groundbreaking conclusion in several ways. Rereading a draft inspired Mezzacappa to frame their work in terms of those lingering questions. Mezzacappa began envisioning what TSI's work meant in light of the questions. Second, he uses abductive reasoning to follow "what if" possibilities out to some fruitful conclusions. If you recall from the prior chapters, abduction is a form of reasoning whereby premises and conclusions share a degree of uncertainty. By conjoining tentative propositions together, he frames a possible world where the instability might contribute to the explosion. He uses a specific premise (that the SAS that they created is unstable) and general premise (that any SAS should be unstable) to tease out conclusions about old projects in light of their new findings. Abduction also occurs in an earlier e-mail that helped the team brainstorm their conclusions.

> If the explosion is indeed "jet-like" (whatever that means), what would the ejecta density distribution look like at late times—i.e.. hundreds of years later as in Cas A? If the polar ejecta was shot out faster then at late times there would be less gas in the polar angles and hence less density? Or is more mass overall sent out the polar directions so that even at late times the mass flux in the poles is greater than the equator?

Here, again, a general (explosions are "jet-like") premise and specific (what ejecta might look like in cases like Cas A) premise are linked together. However, unlike in induction or deduction, in this case of abduction, neither the general nor specific premises have been proven beyond the shadow of a doubt. And, these contingent premises are used to make hypothetical conclusions. Jone Rymer, Carol Berkenkotter, and others have claimed that scientists do come to conclusions about their results while writing. This case exposes how such new conclusions come to pass. In order to shape simulations, devil's advocate and other counterargument techniques dictate simulation parameters. Also, conclusions spring from abductive reasoning in e-mails between Blondin and Mezzacappa. Many of these very ideas—even some verbatim wording—go directly into the second and third drafts. E-mail not only facilitates collaborative writing, it is often the very substance of collaborative writing. Similar wording emerges in e-mails, drafts, and other documents. In many of the instances where the team adds new data to a draft, the information closely resembles conversations they held via e-mail. While I could not capture other informal conversations between team members— such as telephone calls, for example—it would not be surprising if the SAS article borrowed language from those conversations, as well. E-mail conversations chained together contingent generalizations from which the team drew tentative conclusions to retest, then report.

Counterargument, recursivity, and abduction played an essential part in the SAS article, as the team firmed up claims and drew conclusions all the way through the drafting process. In effect, the team worked forward and backward, building on work from the past, moving from e-mail to draft, running additional simulations to substantiate new hunches, checking the warrants that the new conclusion implied, and editing the SAS article to include the news. In this way, the SAS article emerges from multiple, lived experiences (including the e-mail chats and phone conversations). Margaret Syverson describes how emergence is vital to how ecologies of composition evolve (10). Syverson argues that the composition process, and thus the process of making and explaining meaning, is distributed across a complex system of interactions between material, technological, social, psychological, spatial, and temporal realities or dimensions (7–20, 203–206). In the case of the SAS team, words, ideas, sentences, and concepts emerge from the e-mails and resurface not only in the SAS article, but also in other media where the TSI team reports its work. Also, the conclusions emerge not only from the "lab work" programming the software and writing the reports. They also develop from reflection on current, prior and other's work, from abduction, and from counterargument.

THE TEAM RELATES THE FINDINGS
TO THE SCIENTIFIC COMMUNITY

Conversations between fellow SAS team members affected the scope and shape of the article and the argument it makes. Presentations to the segment of the scientific community who comprised the audience at professional conferences also impacted how and when the team framed its findings. The TSI team reported their findings in the *Astrophysical Journal* or *ApJ*, a specialized publication geared toward an academic community. However, the team also sought more wide distribution of their findings. The team had already collaborated on the SAS project prior to its publication, *vis-à-vis* conference presentations and abstracts. The SAS article and drafts were but one outlet for scientific deliberation and activity. E-mail conversations between Blondin and Mezzacappa revealed how presentations prior to publication motivated the team to want to publish immediately, for the sake of priority of discovery. Blondin writes in the first e-mail correspondence, "I would like to put this on the fast track, since we have been talking in the community about this for a while."

The team participated in a few venues at which the team promoted its work prior to publication—some of which were conference presentations. By January 2002—five months prior to the first recorded draft of the SAS article—the group had presented versions of their work at the American Astronomical Society (AAS) 199th meeting in Washington, DC. They titled the presentation "Sources of Turbulence in Core Collapse Supernovae." Their abstract limits the implications of the team's results. In the abstract from the conference, they tout the potential benefits for supernova modeling: "These results have implications for future two- and three-dimensional supernova models." It also states a more general conclusion, *sans* the exclamatory revelation of an alternative mechanism for supernova explosions. My bold for emphasis: "This feedback leads to **vigorous turbulence** even though the postshock entropy gradient is marginally stable to convection." The group also made a similar presentation later at the 2003 High Energy Astrophysics Division (HEAD) meeting—"The Inherent Asymmetry of Core-Collapse Supernovae"—a month after the SAS article appeared in the *Astrophysical Journal*. Small word changes between the two abstracts speak volumes about (1) how reserved the original abstract was and (2) how far the team had drawn conclusions in the later version. My bold for emphasis: "The result is an **expanding, aspherical blastwave** with postshock flow dominated by low-order modes." The two abstracts share some similarities. They are both informative abstracts, in that they both describe the results of the research. Also, they both contain nearly the same content organization—both describe what the team set out to examine and what results the simulation yielded. However, the team draws more narrowly focused conclusions during this presentation session. Notice that, in this abstract, what was once described as a "vigorous turbulance" has now become an "aspherical blastwave" (which details the wave). Here, the team emphasizes the presence of the instability and its asymmetry by making it the main object of the topic sentence. The more time the team spent honing the simulation, the more precise a picture they painted of its significance. However, the first work provides a solid foundation for future work—both the SAS article itself and the next presentation.

The presentations give us points of comparison with which to measure what realities constrain the TSI work. The team frames its work at the forefront of research into computer modeling as well as explanations of supernova explosions. In the presentations, information about supernova isn't the only benefit from TSI work. The team whets its audience's appetites with the prospect of much-needed 3-D modeling. Other genre—presentations and abstracts, in particular—serve as forerunners to the SAS article; they provide much of the foundational data for the paper. The team no doubt also incorporated lessons learned from interactions with peers at the conference. E-mail conversations help the TSI team manage and develop their findings across multiple locations, data from previous, related documents and projects also establish a framework for the SAS article. However, abstracts and slides from a prior conference show that the TSI team began shaping the story of this research long before the first draft of the SAS article.

THE SCIENTIFIC COMMUNITY RELATES THE FINDINGS

The broader scientific audience—those who have read and cited the SAS article—continue to impact the meaning of the work. Citations show that the team's original findings were adopted not only on their own grounds, but also in terms of their usefulness in helping advance other researchers' projects. TSI's rhetorical justifications, though deliberately crafted, were not what lingered in the minds of those who cited the work after its publication. Also, given the theoretical nature of the TSI simulation results, one would expect those citations to distance themselves (via hedging or citation type). In fact, of the over 350 citations referring to the SAS article so far, the reports and articles more or less embrace the SAS findings.

First of all, the citations tend to use the SAS article to advance their own ideas. John Swales outlines the rhetorical moves that most research articles make in their introductions: (1) establish a territory, (2) establish a niche and (3) occupy the niche (141). Authors report the current state of their field; then they find an opening within it and discuss how their current work fits the bill. In the first move, scientific authors can show the centrality of their work, make generalizations about a particular research topic, or review other research in the field. While Swales focused on these moves in research article introductions, I found that they occurred elsewhere in articles that cited the SAS article. For example, Swales's moves appeared in latter sections of literature reviews and year-end recaps.

Many of the articles that cite the SAS article reported the instability for the purposes of mapping out a territory of research or explaining why their own projects turned out as they did, as opposed to contrasting their work from the SAS article. For example, Balantekin refers to the SAS shock in their introduction to explain other aspects of supernova neutrino-nuclear astrophysics; the SAS article is their second reference. "At issue is the fate of this 'bounce' shock (see, e.g., [2]). . . . As the shock transits material beyond the inner core, most of its kinetic energy is dissipated in the photo-dissociation of nuclei. The shock quickly . . . evolves to become a standing accretion shock" (Balantekin et al. 2513). Balantekin starts from the assumption that the accretion or "bounce" shock reported in the SAS article exists. Further, Flanagan et al. cites the SAS article to explain observational data they collected from the x-ray spectrum of an actual supernova explosion: "Two-dimensional models of standing accretion shocks in core-collapse supernovae (Blondin, Mezzacappa, and DeMarino 2003) are unstable to small perturbations to a spherical shock front and result in a bipolar accretion shock" (Flanagan et al. 245). These two citations are exemplary of the majority of citations in that they do not hedge on making claims about the virtual evidence, despite the groundbreaking nature of the SAS article findings. Fewer of the citations used hedges of any kind when reporting the SAS findings than those that did use hedging. For example, Flanagan et al. (above) does not say that the shocks *might be* unstable to small perturbations; it claims that the shocks are unstable, outright.

Furthermore, articles that cite the SAS article interested themselves in the results more so than the methods. The citations referred to the shock rather than the simulation or its methods. Also, citations deemphasized authorial activity and emphasized the SAS article findings. Swales maps citations along two axes: reporting and integration. Reporting citations use "a 'reporting' verb (e.g., show, establish, claim, etc.) to introduce previous researchers and their findings," and non-reporting (NR) citations do not (Swales 150). An integrated citation "is one in which the name of the researcher occurs in the actual citing sentence as some sentence element" (Swales 148). Non-integrated citations acknowledge the work without attributing it to a specific scientist or team. Citations carry reporting and integration qualities simultaneously. Consider the difference between the following statements about the same study: (1) The study shows that disparity widened. (2) Jones et al. found that the disparity widened. (3) Per Jones et al., the disparity widened. And (4) the disparity widened. The first example is reported, but non-integrated. The second is reported and integrated. The third is not reported, but integrated. And the fourth is neither reported nor integrated. Reported and integrated citations (which both name the author and report his efforts within the body of the sentence) work the hardest to give credit to other authors. They draw the readers' attention to the human element and, in turn, the possibility of error. Non-reported and non-integrated citations immediately put the findings and phenomena into the spotlight for analysis, underplaying the author and/or his or her methods.

Writers can integrate the names of researchers into the sentence themselves or leave that information in the endnote or parenthetical citation (which is non-integrated). They can also report the actions of the researcher, or talk about the phenomena without mentioning methods at all (non-reported). Whichever they choose, they do so for different ends: Sometimes they do so to help identify with a band of likeminded colleagues. Other times they serve to place blame or spread the brunt of criticism. Either way, integrated citations attribute agency to people rather than events or phenomena. Non-integrated and non-reported citations foreground findings and phenomena. Reporting the methods and integrating authors' names help qualify a writer's affiliation with the cited work and buffer audience disapproval.

Most SAS article citations were non-integrated or non-reporting. For example, Cardall neither integrates nor hedges the SAS findings in their brief history of neutrino radiation hydrodynamics—bold indicates the SAS citation: "This achievement [of low-mode instabilities] was presaged by earlier studies demonstrating the tendency for convective cells to merge to the lowest order allowed by the spatial domain [45] and recognizing a new spherical accretion shock instability [46]" (Cardall 299). In summary, most of those who cite TSI's work report the phenomena over the methods or the TSI researchers. Under half of the citations hedge the claims with qualifying words such as "might, "may," or "possible." And most neither integrated nor reported the TSI authors or their methods. Authorial agency is a secondary concern for the wider scientific audience; the relevance of SAS article conclusions to the field mattered more. The scientific community trusts the review process to try their fellow scientists by fire. The news reported in the SAS article traveled fast and with minimal opposition. Surely, the rather seamless reception of the SAS article is a testament to the TSI team's thorough research and precise work. It also bespeaks the efforts made by the TSI team to predict and redress audience questions and concerns in their work. Making presentations and playing devil's advocate prior to submitting the article helped strengthen the argument and render it more acceptable to the scientific community. Furthermore, the news of their findings had use, insofar as the SAS findings helped other researchers understand and explain their own findings and put together the pieces of their own work. Most of the citations used the SAS article to explain their own position or findings. Those who cited the SAS article believed the story. The simulation contained virtuous and powerful suppositions; the strength of the simulation lies in its explanatory power.

The TSI team's deliberations (in e-mails about the manuscript, in prior draft of the manuscript, in conferences presentations about their findings) and the wider scientific community's reaction to the SAS article all contribute to the relational meaning of the SAS article. These documents and conversations make transparent the relationships between evidence and decision making and past information and new claims, and interactions between material, social, and temporal factors. Audience expectations and needs play a major part in making what the SAS article means. The team revisits old projects and anticipated audience concerns in order to interpret the SAS project. As they enact the writing process, conclusions emerge. Then, the wider scientific community parses from the SAS article what is useful to their own endeavors. The SAS project lives on not as a complex knot of rhetorical moves or the ad hoc decisions of well-trained scientists, but as a nugget of information—"news"—about how supernovae explode. What is it that makes an idea catch fire and circulate over others? I argue utility, explanatory power, and virtual vividness or potential. Instability in accretion shocks may very well supplant or complement the neutrino theory about how supernovae explode. For now, other scientists use it to help explain their own phenomena, make sense of lingering questions and justify the importance of their own work.

THE TEAM RELATES FINDINGS TO THE PUBLIC

The scientific community informs how and when the team framed its findings; then, it distributes versions of the story of the SAS project—versions
that help advance their own scientific inquiries. Reports of the SAS project to
the general public, on the other hand, did not affect the substance or direction
of the SAS findings directly, but rather the general public solicited different
pictures of the potential and residual impact of the findings. It is in public
reporting of the SAS project where we see that messaging about the project
could affect funding and the future of energy production in the U.S. Peripheral documents demonstrate how governmental interests contextualize the SAS
findings. The team framed its work differently to reflect the different social
interests of their public audiences. For example, when the Department of
Energy describes TSI's work in newsletters and annual reports, the focus
shifts from explaining a supernova mechanism to practical applications of the
technology used to create the simulations. Mezzacappa explained in an interview prior to publication of the SAS article: "Our work addresses very broad
themes important to the DOE's national mission . . . the ability to model the
movement of radiation through matter and its interaction with that matter is a
concern . . . for people who model internal combustion engines, climate
patterns, and . . . better tools for radiation therapy" (Walli). The team tailored
its message about the DOE in such a way that justifies spending funds on the
TSI projects. Governmental affiliations affect what TSI claimed was the
import of their work. A year after the SAS publication, Mezzacappa reported
the news—that they'd discovered instability—in a news brief for Krell Institute, a company that helps support many government-sponsored science and
technology projects. The article describes the team and their work: "[T]his
multi-disciplinary team approach is paying off. . . . 'We discovered that the
shock wave itself can become unstable, either aiding or altering the shape of
the explosion,' says Mezzacappa. 'Like the SciDAC name says, that was
scientific discovery through advanced computing'" (Krell Institute). Here,
again, what TSI does, and what they found, is reported as closely aligned
with two funding agencies—namely, SciDAC's purpose and the Krell Institute's interests.

In turn, the organizations that sponsor the TSI team cite the SAS project
in their annual reports. Governmental agencies value such annual reports.
They use the reports to justify funding, report to the Congress and others for
whom it is important to keep an accounting of how they spend taxpayers'
dollars. For example, in 2003, editors and scientists prepared an annual report to the director of the Office of Science of the Department of Energy. The
transmittal letter articulated the importance of the report.

A major increase in investment in computational modeling and simulation is appropriate at this time, so that our citizens are the first to benefit from particular new fruits of scientific simulation, and indeed, from an evolving culture of simulation science. . . . Based on the encouraging first two years of the Scientific Discovery through Advanced Computing initiative, we believe that balanced investment in scientific applications, applied mathematics, and computer science, with cross-accountabilities between these constituents, is a program structure worthy of extension. (Colella et al. 1)

Annual reports like this one explain and legitimate scientific research to policy makers and granting bodies. The report editors above cite SciDAC's work, in particular, of which TSI is a subgroup. Later in the report, the editors discuss TSI's results as "newly discovered shock wave instability and resulting stellar explosions (supernovae), in two- and three-dimensional simulations" that are important to the Office of Science insofar as (1) "core collapse supernovae . . . are the dominant source of elements in the Periodic Table between oxygen and iron" and (2) "extremes of density, temperature, and composition . . . in supernovae . . . [involve] . . . fundamental nuclear and particle physics that would otherwise be inaccessible in terrestrial experiments" (Office of Science 29). Report editors have these tangible benefits in mind when, in the transmittal letter, they mention the "fruits of scientific simulation" worthy of more government funding. The TSI simulation scientists are leaders in their field. The team also underscores the broader value of simulation work, which, on its own without filtering, might seem rather abstract to some. Here, the report editors promise both "fundamental and practical" benefits, but also advantages for fuel-consuming citizens. Essentially, annual reports work like marketing materials for organizations like SciDAC and the DOE Office of Science. Report editors explicitly align their scientific objectives with administrative (if not political) ones.

Another annual report, published by the National Energy Research Scientific Computing (NERSC) Center in 2001, offered some of the conclusions ultimately made in the SAS article. NERSC hosted one of the larger labs in the SciDAC affiliate. Report editors could not announce the instability outright, because TSI had not drawn that conclusion yet: "[T]he blue and green regions depict the turbulent environment beneath the supernova shock wave" (NERSC 73). Here, it is no longer instability, but an undefined "turbulent environment" in question, whose answer will provide a "link in our chain of origins from the Big Bang to the present" and "a new class of multidimensional supernova models" (NERSC 73). The turbulence in and of itself does not justify TSI's work; the possibility that understanding the instability might lead to discovering new means for binding energy does. The latter goal more closely aligns with DOE explicit directives. Annual report editors repackage and reframe the value of piecemeal results.

Other points of comparison further evidence the team's attention to institutional affiliations. The team gives a nod in the article to their funding sources. In the very first e-mail transmittals, Mezzacappa reminded Blondin to add an acknowledgment section to the article: "I have a boiler plate that has to go in the acknowledgments at the end, plus I want to site SciDAC too. We can add these when we're near a final draft. I will e-mail the boiler plate to you for inclusion." Within the drafts that Blondin sends to Mezzacappa, Mezzacappa mentions to Blondin where to add SciDAC's name to a long list of acknowledgments including Oak Ridge National Laboratory, North Carolina Supercomputing Center, and the Centers for Computational Sciences. Gross and Harmon reported how most of twentieth century scientific articles included acknowledgments of funding agencies, and other scientists who helped the research (181). Gross, Harmon, and Reidy argue that boilerplates and other such acknowledgments originated out of the "growing complexity of work arrangements" between small groups of scientists, large lab affiliations and funding agencies (Gross, Harmon, and Reidy 181).

Furthermore, since external funding affords simulation resources, it has potential to constrain the simulation process itself. In the second round of e-mail transmittals (July 2002), Blondin reports a computer crash that delayed completion of other simulation iterations. "Things were going well on the 3-D simulation until our IBM came crashing down. . . . I am having severe problems with the speed of writing . . . , but [we're] working on it. Once that is fixed I can continue . . . if our IBM is not resurrected soon." The crash raises a dilemma—Blondin wants to take their findings to a conference. "I have 190 3-D datasets on disk, and asked Ross if he can make a nice 3-D rendered movie next week that we can take to our respective conferences." Had TSI more funding to buy more capacious, fast or expensive computers, would it have delayed them? Did knowing the limitations of their own system impose further limitations on the kinds of ad hoc decisions they could and could not make?

Software release timing and revisions do affect the freedom with which scientists share code. Mezzacappa explained how the newest versions of simulation software haven't been released yet for copyright reasons. In the long run, TSI will distribute the software, as they did older generations of VH1. "While some groups under SciDAC, like Mezzacappa's, are developing the next generation of community codes for users, other groups are building tools for the developers, themselves, so that the latest algorithmic technology migrates into not just one application, but is available across a common interface for many applications" (Keyes). The code itself becomes its own publication, scholarship, and justification for reappointment and additional funding. Long before and long after the SAS article was published, TSI jockeyed their research for institutional and scientific advancement and laurels. Article writing and simulation decision making start not at the beginning of a project, but well beforehand in presentations, grants, and other peripheral documentation and at the mercy of funding agencies, media, resources, and time.

DISCUSSION

The SAS article uses rhetorical figures to put forth an explanation of what findings from the SAS project mean. The rhetorical figures used to deliberate and draft the report, in turn, show how that meaning was constructed with physical evidence, counterarguments, decision making, weighing alternatives, and other careful deliberations in mind. Interpersonal communication between team members and between the team and conference goers affected the shape and expedience of the findings. And additional, external communiqués from researchers, organizations, and public relations representatives citing the SAS project reported the most compelling aspects of the SAS project for distribution and dissemination. The meaning of the SAS project entails not only the published article, but also recursive deliberations, prior conference presentations, and anticipation about organizational aims and audience expectations and needs. The physical limitations of simulation resources and organizational funding might have inspired particular deliberative rhetorical figures in earlier drafts of the SAS article. Time and hardware constraints might have narrowed options for choosing symmetries and running iterations. Furthermore, that the team anticipated audience objections in e-mail conversations might have been why they did not report more deliberation and rebuttals for explaining their ad hoc choices in the published work. They might have felt that the text already answered those questions and that they answered conference audiences questions' in prior venues. Interactions with peers during conferences (where they began reporting the SAS project long before the article) might have made counterarguments and other audience objections more vivid during drafting. In fact, given the manner in which others cited the SAS article, it's unclear that the actual audience critiqued the paper as much as the team itself did its own drafts during recursive e-mail exchanges.

The published SAS article might have underplayed the team's agency, but the document trail of prior drafts, e-mails, presentations, newsletter articles and other writings about the SAS simulation revealed how counterargument, others' work, prior TSI articles, deliberation and other rhetorical figures justified their ad hoc decisions (such as setting parameters, making eliminations, and adding creative elements to the simulation). In fact, during the drafting process, the team developed the most important finding in the SAS article—the "news" about an alternative mechanism for supernova explosion—thanks in part to abduction and creative fervor. However, those who cited the SAS article in their work paid modest attention to the intricate rhetorical process; instead, most of the citations repeated the "news" from the SAS article for their own purposes. We also saw how social dimensions (from governmental institutions and the scholarly community) might have impacted not only how the TSI team framed and reported their findings, but also whether or not they had the resources for making and sharing simulation software. It appears that the team reframed the story about the SAS project to emphasize its usefulness for newsletters and other organizational records.

The SAS team retained rhetorical consciousness throughout the drafting process and as they tailor the news about their simulation from venue to venue. This finding runs contrary to some studies in the rhetoric of science and technical communication where scientists are often construed as either hostile to the rhetorical process or ignorant of it. Take, for example, what Berkenkotter and Huckin discovered about June Davis, a biologist encouraged by journal editors to show how her work on a particular fungus pertained to a substance that can improve the body's natural response to diseases such as cancer. Davis, like most other scientists, saw her "laboratory research and rhetorical activity as distinctly *separate*," and she thought that "contextualizing" and "recounting" the events is tantamount to telling a "phony story" (Berkenkotter and Huckin "You Are What You Cite" 58). Davis all but resented having to alter her introduction so as to connect her work (even remotely) to a larger issue of concern. Compare Davis to the TSI team, who substantiated some ad hoc decisions and eliminated others from later drafts of the article. They spoke differently about the same research in the newsletter than they did in the article. The TSI team understood early on that computational processes and lab activities aren't above rhetorical or social implications.

New studies in the rhetoric of science have revealed how the drafting process births scientific conclusions. My textual analysis showed how abduction, rhetorical figures, and counterargument techniques helped TSI set simulation parameters. In the case of the SAS project, abduction and counterargument—often thought to play an integral role in hypothesis making—helped inspire new conclusions during the drafting process. Furthermore, according to Fahnestock and Secor, science typically enlists the lower stases (i.e., existence, definition) in writing and argumentation. However, this chapter shows how ad hoc decision making involves higher stasis (i.e., cause, evaluation, and proposal) for evaluating alternatives and justifying reasons for making decisions.

In comparison, Jone Rymer used interviews and compose-aloud protocols and found two ways that scientists develop findings: (1) For some, the research article directs activities in the lab and, (2) for others, the writing process begets the research article (Rymer 238, 239). My research uncovered another type: scientists for whom the creative process begins well before the first draft (in previous incarnations of ideas such as conference presentations) and extends well beyond the article drafting process (in abstracts, presentations, and media coverage). Article writing began for TSI not at the beginning of a project but well before, in presentations, newsletter interviews and other peripheral documentation. In the case of the SAS article, I showed how computational studies in astrophysics build from pre-texts and drafts. This runs contrary to many other studies that limit meaning production to the drafting process. The "point" of the SAS article evolves from pre-texts, recursive drafts, and excitement about the possibilities. In the case of the SAS project, the production of meaning also extends into how other scientists and the general public interpret the simulation findings.

The SAS case also has implications for prior studies on the reading habits of scientists. To date, no study exists of the reading habits of astrophysicists in particular, but Charles Bazerman does discuss how physicists read physics. Readers skim for information directly pertinent to their own interests and chuck the rest. Often, they read methods sections for no other reason but to look for new methodological tricks. They also call on personally organized knowledge (schema) for scanning (words, names of phenomena and researchers) and predicting content (based on the cue words) (Bazerman, *Shaping*). Then they use this information to put the article within their personal schema of the field. This map of the field is shaped by problems in the field, personal perspectives on the advancement of field, and the readers' own set of research. Considering how physicists read texts, how the TSI team excluded details about their methods from the published draft, and how other researchers' cited the most important claim in the SAS article without much alteration or hedging, it may be the case that the scientific community should consider developing new formal procedures or genres for sharing more details about the decision making that underpins simulation science.

Ultimately, given what we've learned from all of the data, five criteria governed what ad hoc decisions they chose and how they defended them: (1) *They were quantifiable.* The TSI team chose some methods because those methods produced quantifiable data to which to compare. (2) *They were physical.* The team took key components of physical evidence to duplicate in the simulations—key insofar as they were basic features established by observational data. (3) *They were comparable.* The team chose conditions and methods specifically for comparing their work to others. (4) *They were reasonable—they had internal logical.* The team set parameters justified by their own work's ends. They wanted the simulation to test and achieve certain results—and avoid disruption from unwanted elements. This desire often decided parameters and ad hoc actions. (5) *They were believable.* The team used abduction and counterarguments to determine which decisions their audience would have a tough time believing and which ones their audience would accept intuitively (or, at least, without reservation or question).

Chapter Six

The Rhetoric of Simulation in the News

It is important to understand simulations as rhetorical products for many reasons. First, simulations are virtual evidence. In other words, they represent the virtue (or essence) of an event or phenomenon that scientists cannot easily confirm or verify with direct observation, even in cases where they fail to embody every detail of the actual phenomena with the utmost accuracy. In this way, they have *energeia*, or they are in the gradual and contentious process of bringing before our senses the potency and virtue of the actual thing. They bring before our eyes the forming of relationships between extra-linguistic and linguistic events; they reflect not only what we know, but also what we are coming to know and understand. Second, simulations are best understood as having relational meaning; in order to understand them, we must make conspicuous the relationships between linguistic and extra-linguistic phenomena and conditions that determine their contextual value. The book thus far has argued that strategies of rhetoric and argumentation such as abduction, narrative, and others reveal the relationships that underpin the construction, reporting and distribution of a simulation's meaning. This chapter will explore how this framework can shed light on simulations in current events and the news.

SIMULATIONS AND CLIMATE SKEPTICISM

Understanding the rhetorical nature of simulations can help illuminate the global climate debate. Recent polls suggest that a large percentage of the American public is skeptical about global warming (Saad). In contrast, the scientific community has come to a fairly wide consensus affirming the existence of the phenomena (Anderegg et al.). How is it that, despite the scientific community's widespread certainty, climate skepticism persists and thrives in the general public? Part of the discrepancy has to do with the fact that simulations are how climate scientists derived their consensus around the subject.

The Intergovernmental Panel on Climate Change (IPCC) report is one of the most complete articulations of evidence supporting the existence of global warming. The report is comprised of several sub-reports that synthesize many layers of data vis-à-vis simulations that process hundreds of data points. The virtual evidence produced by the simulations ultimately suggests that atypical climate events indicate a global warming trend. To construct the simulations (i.e., synthesize the data and produce virtual evidence), the IPCC report uses rhetorical conventions similar to the ones outlined in prior chapters. The 2007 sub-reports, for example, include a detailed explanation of how IPCC climate models are constructed. Their methods sections explain how they handle uncertainty within the parameters of their models. For example, they explain how, for parameters that cannot be measured, they perform a practice common to simulations of adjusting "parameter values (possibly chosen from some prior distribution) in order to optimize model simulation of particular variables or to improve global heat balance. This process is often known as 'tuning'" (IPCC Contribution of Working Group I 8.1.3). They also account for the inevitability that their models include "significant errors" (IPCC Contribution of Working Group I 8.1). They explain that "limitations in computing power," "limitations in scientific understanding or the availability of detailed observations of some physical processes," and "models that "display a substantial range of global temperature change" do not dismiss the fact that, despite the uncertainties, "models are unanimous in their prediction of substantial climate warming under greenhouse gas increases, and the warming is of a magnitude consistent with independent estimates derived from other sources" (IPCC Contribution of Working Group I 8.1).

In particular, the report contains several sections that outline IPCC methodology, including a system by which they value and rank the virtual evidence that they use to mount their analysis. The report from the Working Group III on mitigating climate change, for example, includes an entire section (2.3) that describes how its model factored in uncertainty in order to produce findings in such a way that minimizes bias and overemphasis on uncertain or unclear information. The group determined the level of agreement on any given finding factored into the IPCC models. Findings with low levels of agreement and limited amounts of evidence were weighed less heavily than those findings with high agreement and lots of evidence. The IPCC describes how likelihood and levels of confidence are deemed high only where there is " high agreement and much evidence, such as converging results from a number of controlled field experiments" (IPCC Contribution of Working Group III 2.3.1).

Rating scales such as the one used by the IPCC are often used in evidence-based scientific syntheses. For example, in medical science where studies and findings proliferate, organizations such as the Agency for Healthcare Research (AHRQ) and Quality in the U.S. and the Cochrane Collaboration in the United Kingdom prepare systematic reviews and meta-analyses of randomized controlled trials and other studies on the effectiveness and safety of drugs. Each of these agencies produces executive summaries and other products for clinicians, consumers, and policy makers based on the findings from hundreds of research studies on particular medications, treatments and interventions for particular medical conditions. In both cases, the organizations do not treat all evidence as equal. In the hierarchy of evidence-based medicine, for example, findings from large randomized controlled trials (RCTs) weigh more and matter more than findings from interview or focus group studies because the sample size and methodology of RCTs yield more generalizable findings (Owens et al.; Cochrane Collaboration). Therefore, AHRQ and Cochrane devised strength of evidence scales from low to high with regards to the findings that they synthesize from multiple sources. In turn, strength of evidence scales and other quality assessment systems help signal to the scientific community the strength or certainty of the claims made. They help the medical community estimate the strength of the findings. The IPCC uses their evidence scale in ways that, per rhetorical convention, are meant to address counterargument. These counterargument strategies, in turn, usually build credibility for the argument. They also follow a common virtue of scientific rhetoric—namely, they provide sufficient methodological details to afford others the ability to replicate and validate the findings.

On the contrary, climate skeptics turn these counterargument strategies against the IPCC to discredit their findings. They contend that any level of uncertainty renders void the findings of the IPCC report altogether; in essence, they exploit the traditional courtesies extended in scientific reports of acknowledging limitations of methods to challenge the validity of arguments affirming global warming. Nowhere is this more apparent than in the report from the Global Warming Petition Project, an initiative lead by over 31,000 scientists—over 9,000 of whom with PhDs, per the website's home page—who signed a petition refuting global warming and the IPCC findings. The group published a synthesis review, which is, to date, one of the most comprehensive syntheses of evidence refuting global warming.

The report, entitled "Environmental Effects of Increased Atmospheric Carbon Dioxide" essentially sets out to discredit affirmative evidence in the IPCC report. In this synthesis review, the argument contesting global warming differs significantly from the IPCC report. First, their review of counter-evidence lacks a clear statement of methods. We, the readers, are given no indication of how the researchers accounted for uncertainties in their own estimation of the data. We are also not told the extent to which certainty was reached by consensus or the strength or weakness of the counter-evidence presented, nor are we told how evidence was weighed to reach the conclusions that the review draws. The report only indicates that the team who completed this review also completed a prior review: "When we reviewed this subject in 1998 (1.2), existing satellite records were short and were centered on a period of changing intermediate temperature trends" (Soon et al. 1). Unlike the IPCC report which shares many details of its methods in attempts to make its methodology transparent and verifiable or falsifiable, the "Environmental Effects" report does not describe the methods it uses to synthesize data and reach its conclusions. Second, also unlike the IPCC report, the "Environmental Effects" review does not incorporate several, layered rhetorical strategies in order to make different points or accommodate different stages of decision making or different degrees of certainty. To challenge some of the claims of the IPCC, it primarily (if not exclusively) uses refutation by offering up counter-evidence and decrying weak evidence. For example, the "Environmental Effects" report calls attention to the fact that the phenomena underpinning the greenhouse effect are "poorly understood" (Soon et al. 2). It presents figures derived from a few counter-studies to show how "the change is slight" in surface temperatures in the United States (Soon et al. 3). It reports how one study shows that "surface temperatures and world hydrocarbon use are not correlated" (Soon et al. 3). The title for one of its figures argues that "There Has Been No Increase in Number of Atlantic Hurricanes That Make Landfall" (Soon et al. 3). It also prioritizes experimental data, using it to refute simulation data pointing to climate increase: "Does a catastrophic amplification . . . lie ahead? There are no experimental data that suggest this. There is also no experimentally validated theoretical evidence of such an amplification" (Soon et al. 3). In doing so, the report privileges laboratory and experimental evidence and findings without acknowledging that (1) the IPCC report bases its claims upon syntheses of several pieces of experimental data and (2) experimental data are not without degrees of contrivance and uncertainty, particularly when extrapolating findings from an experiment to the world, at large—a world without controls and with several confounders and interrelated variables at play. One representative excerpt from the report epitomizes its objective—namely, to show non-causality: "Correlation does not prove causality, but non-correlation proves non-causality" (Soon et al. 5). In essence, the report offers up as many studies as possible refuting increases in atmospheric and surface temperatures and the impact of carbon dioxide.

Ironically, the "Environmental Effects" report uses projected estimates and ad hoc reasoning itself to advocate for nuclear energy, which could be construed as politically motivated or leaning toward special interests invested in nuclear technology. It uses if-then logic to make claims about a projected, future efficiency of nuclear power that is not currently a reality: "If just one station like Palo Verde were built in each of the 50 states . . . , these plants . . . would produce 560 GWe of electricity" (Soon et al. 11). Also ironically, it relies on longitudinal data about the climate to make its refutations, then it dismisses the reliability of similar longitudinal data used in IPCC climate models. The report says that the models supporting climate change "have substantial uncertainties and are markedly unreliable" which is unsurprising because the climate "is very complex" (Soon et al. 7). In terms of argumentation, the IPCC uses several rhetorical strategies to make decisions that factor in the complexity. In contrast, the "Environmental Effects" synthesis only points out the complexity without explaining how their analysis accounts for it, or how the complexity might render their own calculations incorrect or unreliable. Overall, the report does not offer up as much evidence as the IPCC report—123 citations compared to well over five times as many in the IPCC sub-reports. The citations are also older than those in the IPCC report—about half of the citations in the "Environmental Effects" report were from the 1990s compared with the overwhelming majority published in the twenty-first century in the IPCC reference lists. The tradition in the production of scientific knowledge is that newer citations are more accurate insofar as they build on older knowledge and revise, as necessary. Overall, the "Environmental Effects" report does not reveal its methodology, nor does it consider or explain the strength of the evidence it offers up.

There is also a theoretical irony underpinning the difference between the reports. Politically speaking, climate skeptics are the conservatives in the debate. Their reservations about global warming science are shared by those opposing policy makers who would use climate science to create and enforce what they see as unnecessary regulations on industries to curb the production of greenhouse gases. They frame their argument as such. Political optics pit global warming scientists with progressives who advocate for new and green energy sources. Progressives cite the climate trends reported in the IPCC report to campaign for political action. However, from the perspective of the rhetoric of science, the IPCC report is the more conservative report, insofar as it abides by rhetorical conventions of the scientific community. It makes concessions and provides the necessary methodological details to support the validity, falsifiability, and replicability of their findings. In contrast, the "Environmental Effects" report deploys inconsistent rhetorical maneuvers. Insofar as the report asserts its claims without considering or revealing the degree of certainty of the evidence it used or any other methods it used, it resembles positivism, or assertion of claims without self-reflection. However, insofar as its primarily modus operandi is to refute claims to the affirmative, it resembles deconstruction or drawing distinctions, dismantling affirmative definitions, and teasing out the contrary points, or *différance*, that underpins claims. To this extent, climate skeptics use postmodern argumentation with roots in relativism in order to assert conservative interests and refute climate change.

As I mentioned in an earlier chapter, climate skeptics point to Climategate as proof that global warming science is unfounded and untrue. In particular, email exchanges between Climate Research Unit (CRU) team members that describe how they derived their findings were accused of containing admissions of falsification and evidence suppression. Similarly, they accuse the IPCC of doctoring the data to produce findings consistent with their political agenda. The 2007 IPCC report faced intense criticism over claims made regarding the melting of Arctic sea ice, the effect of melting ice sheets, projections about the date of Himalayan glacier melt, and a claim about African crop yield. Some felt that the report underestimated the threat of some of these issues. However, climate skeptics felt that the IPCC report overestimated the threat, and that it did not base its claim regarding crop yield on peer-reviewed evidence. In response to the uproar regarding the IPCC report, several scientists wrote an open letter reminding the public that the IPCC report only included a "handful of mis-statements (out of hundreds and hundreds of unchallenged statements)" and that none of those misstatements "undermines the conclusion that 'warming of the climate system is unequivocal'" (Piltz). Overall, the IPCC methodology offers a subtle argument that admits weaknesses and attempts to account for them in how it factors in data into its climate simulations. Ironically, the counterarguments lobbed against them fail to report—and possibly fail to use—the same methodological rigor and care.

If we consider these controversies in light of the arguments and rhetorical strategies necessary for creating virtual evidence, they lose some of their fire. First, consider the CRU e-mails in light of the rhetorical deliberative process that simulations necessitate—namely, the deliberative process that prior chapters in this book outline. In this new light, the Climate Research Unit made no malicious deletions or revisions, but rather, it made rhetorical calculations to justify algorithmic ones. For example, one of the e-mails between Climate Research Unit scientists admitted that for one sub-routine of several investigating energy flows in short-term climate variability: "The fact is that we can't account for the lack of warming at the moment and it is a travesty that we can't" (Romm). However, popular media misreported this detail as an overall statement regarding global warming in general. That the team could not account for this phenomena represented by one sub-routine does not subtract from the weight of evidence where they could account for the presence of warming, nor does it diminish the overall impact of the virtual evidence—namely, that the simulations produced findings with the potency and virtue of actual climate change that is taking place. Furthermore, the

IPCC report itself admits that the model includes and accounts for uncertainties. That reporters uncovered this admission should have come as no surprise to skeptics or to the media, because the IPCC report itself admits as much. In another email, the Climate Research Unit team described the ad hoc reasoning that it used to account for a well-known problem in climate science called the tree ring divergence problem (i.e., the disagreement that exists between temperatures measured by thermometers and temperatures measured via interpreting the formation of tree rings) as a "trick" to "hide the decline" in temperatures. Again, in this case, the team described the results of their ad hoc decision making in the context of handling uncertainties of one sub-routine in their simulations, not about the sum total of their simulations' findings. However, the skeptics and the media took the quote out of context, most likely out of a misunderstanding of the legitimate decision making and rhetorical machinations that all simulations require. Recall from prior chapters that Winsberg, Iverson, Mezzacappa, and Z. Jane Wang used similar terminology—tricks, ad hoc, ingenuity, and so forth—to describe the extralinguistic and creative decision making that went into setting parameters and initial conditions of simulations. Using such terminology is not uncommon or nefarious in simulation science; rather, it is a common means by which teams of simulation researchers indicate among themselves where strategies of rhetoric and argument helped the team construct virtual evidence.

These rhetorical machinations do not invalidate the virtue or potency of simulations to explain the world and its phenomena; on the contrary, the most effective simulations acknowledge and account for the uncertainty and gaps, and they construct compelling arguments that help bring phenomenon before our eyes and to our senses. The same accusations of imprecision, obfuscation, cherry-picking could be made of the "Environmental Effects" report because the authors who wrote it made their methods opaque, rather than revealing them as is the courtesy and rule of argument of scientific rhetoric. Comparing the "Environmental Effects" and IPCC reports, we see that the former report failed to consider copious citations (over 5 to 1) and rate the strength of the evidence they offered and the claims that they made.

It is clear from this case that postmodern science is very self-conscious about the fact that its findings may not be the teleological truth. And this new era of scientific rhetoric requires scientists and simulation scientists to be more forthcoming about their methods and the strength and weakness of their evidence and claims. The "Environmental Effects" report is more rhetorically simplistic. It does not build a strong, affirmative case of an argument against climate change, but rather uses primarily one rhetorical figure— namely, *negatio*—to challenge the case for climate change. On the other hand the IPCC report enlists a range of rhetorical devices, including concession and evaluation of the strength of evidence. Nonetheless, while the complexity of the IPCC report is a successful rhetorical strategy for high-information audiences such as peer scientists and global policy makers, that same complexity may confound the general public and, thereby, provide further ammunition to the opposition, who points to the complexity of the argument as a sign of weakness and a reason for doubting its validity.

Others who have written on this topic focus on the contentiousness of the debate, the failings of media coverage, and ways for climate scientists to improve how they present virtual findings. Regarding the debate, Andrew Hoffman's study of the rhetoric surrounding the climate change debate describes how the opposite sides take stances across a logical schism wherein they share no grounds or assumptions and, therefore, seem to debate different issues altogether. Leah Ceccarelli argues that the debate is a "manufactured" scientific controversy insofar as climate skeptics deploy rhetorical traps such as overstating to the public the uncertainty of scientific facts, and misusing appeals to centrism, open-mindedness and fairness in order to undermine the large scientific consensus on the matter ("Manufactured"). Regarding the media coverage, research suggests that journalism's compulsion for localizing global stories, dramatizing stories by using apocalyptic narrative, and providing a compelling story or "hook" has led to misrepresentations and misreporting of climate science and climate change findings to the general public (Hulme; Spoel et al.; Wilson). Susanne Moser describes how the parameters of climate change—the fact that neither the causes nor its most dire implications are immediately visible or manifest, the delayed gratification (if not outright pain and financial impact) of taking action, the associate degrees of complexity and uncertainty, and so forth—complicate communicating findings to the general public. Finally, other scholars have criticized climate scientists themselves for missteps made as public scholars. Critics point to value-laden word choice used to describe their findings to policy makers and unwise and ineffective responses to their opposition (Walsh, "Visual Strategies"; Ceccarelli, "Manufactured"). In particular, Lynda Walsh questions the uncertainty underpinning the IPCC's use of confidence intervals and projections to create further projections, and she criticizes the reports' use of language and visuals reporting their findings that seem to prophesy doom and beg an emotional reaction and moral obligations to act (Walsh, *Scientists*).

Understanding simulations as articulated in this book can help address many of these issues. First, the schisms that separate the sides of this debate are manifest, in part, because they are using different rhetorical strategies. Negating global warming is the primary mode of operation of climate skeptics who wrote and undersigned the synthesis review; consensus building and acknowledging uncertainty, the overarching methodology of the IPCC report. The IPCC reports acknowledge uncertainty; the opposition, on the other hand, wants more admissions of uncertainty, without acknowledging its own methodological weakness. So, the sides actually share the assumption that simulation science includes degrees of uncertainty; however, only one side discloses its methodology and holds it up for inspection. This book also demonstrates how audiences are complicit in transforming what simulation scientists report in public about events versus what we should do about those virtual findings. For example, remember how, in earlier chapters, the SAS team reported the same findings to different media outlets and public forums, and how they responded to the interests of their audience. They did so not to change the simulated facts or virtual evidence, but rather to explain their potential implications or application. Walsh and Ceccarelli, then, make an excellent point that public scientists must take care to explain their methods in plain language and choose words and graphic design carefully. However, the strain of climate skepticisms that would dismiss simulation science for incorporating degrees of uncertainty does both sides of the debate in particular, and science in general, a disservice. To attack simulation science for uncertainty means that we can and should equally scrutinize the uncertainty of evidence used to sustain skepticism. Furthermore, to attack simulation science for uncertainty discounts the usefulness and virtue of uncertain science.

It is true that politics can impact how findings are presented to the public. As we saw in the case of the SAS project as it trickled through the scientific community to the general public, the TSI team reported the SAS project in such a way that aligned with the interests of the general audience with whom they shared the findings. Walsh argues that, in reporting genres aimed at public distribution (such as executive summaries and condensed reports for policy makers), climate scientists have used a style of reporting (i.e., word choice, color choice) that borders on recommending to audiences what they should do rather than reporting what they found ("Before Climategate"). Issues of uncertainty do warrant examining the role of the public scientists— that is, the part that they play in helping set public policy. Also, great care should be taken not to understate or overstate the implications of scientific findings, no matter the interests of the audience. However, the *energeia* or virtual potency of scientific evidence balances the uncertainty, insofar as it provides sufficient information for action despite the tentative, uncertain or incomplete nature of scientific findings. Sometimes scientific, virtual evidence tells a story so compelling that it warrants and justifies action despite uncertainty.

That some climate scientists should have used more careful language or design techniques to convey the message does not diminish the virtue or potency of their findings. Furthermore, unpopular messages may open up climate skeptics to political attacks, derision, and criticism, but the imprecision of some skeptics' syntheses of findings, oversimplicity of their argument, and resistance to reporting methods with clarity or accounting for the uncertainty of their own argument strips their argument of relational meaning and *energeia*. The counterarguments against climate change evidence have already been factored into climate simulation synthesis. The findings emerge in spite of those issues, not ignorant of them. Furthermore, social realities contribute to the meaning of scientific findings and virtual evidence, but they do not disqualify them. Controversies surrounding overstatements should convict climate scientists to take more care when presenting their findings for public consumption. However, indelicacies of presentation should not diminish the usefulness, virtue, or potency of climate change findings. Public reports must avoid perpetuating the misconception that uncertain science and virtual evidence is invalid, suspicious, or unuseful. Contemporary scientists—the authors of the IPCC reports and the synthesis rebutting the report alike—share similar pressures regarding research funding and public accountability. Those realities, however, do not render virtual findings null and void.

SIMULATIONS IN HURRICANE PLANNING

Awareness of the rhetorical nature of computer simulations can also help us understand some of the trouble and lack of preparation that plagued the response to Hurricane Katrina. During the Congressional investigation into what went wrong, the inquiry revealed that emergency agencies had conducted preparation exercises for a Category 3 hurricane before Hurricane Katrina took place. Prior to Hurricane Katrina, researchers from Louisiana State University's Hurricane Center had to design "worst case scenario" simulations predicting what might happen if a Category 3 hurricane landed in New Orleans (U.S. House of Representatives). In July 2004, over a year before Hurricane Katrina ravaged the Gulf Coast, FEMA hired private contractors to conduct preparation exercises in New Orleans to ready FEMA for the event of a large-scale hurricane. Strategic workshops and drills were based on a computer simulation of a slow-moving, Category 3 storm named Hurricane Pam and the catastrophic conditions that would result. To their credit the simulations forecasted several aspects of the devastation with alarming and eerie accuracy, including details regarding death toll and damage to property. The team of emergency management consultants who helped design the exercises also claimed that they wanted to create a realistic exercise (U.S. Senate). Unfortunately, plans for the anticipated rescue, aid or evacuation fell short. Although highest levels of government officials knew about the simulations and the Pam exercises, neither the virtual evidence nor the planning it inspired facilitated rescue efforts (U.S. Senate 27).

Surely external factors such as politics and timing played a part in the failure to implement the lessons learned from the Hurricane Pam exercises. Congressional documents and other reports of the rescue planning and implementation detail how, when FEMA was moved under the umbrella of Home-land Security, it lost much of the funding and budget it used to respond to large-scale disasters (U.S. Senate; CREW; U.S. House of Representatives). Furthermore, it was reported that the proximity of events—Hurricane Pam exercises completed in July 2004 and Hurricane Katrina hit land in August 2005, thirteen months later—did not afford Gulf Coast agencies and respond-ers to fully complete, implement, and operationalize the planning that the Pam exercises made everyone aware needed to take place (U.S. Senate, U.S. House of Representatives). However, how agencies and responsible parties interpreted the simulated hurricane also played a part. Some felt it only produced a bridging document for future fleshing out and implementation (U.S. Senate 50). Others felt that it was the final plan (U.S. House of Repre-sentatives 83). Also, climate scientists involved in creating the simulation were met with some disbelief and false sense of security on the part of the Army Corp of Engineers. They scoffed at the prospect of a levy breach, despite the warning of scientists and other engineers who said it was an inevitability (van Heerden and Bryan; McMaster, Standring, and Stubber-field).

If the point of the Pam exercises was to create a realistic picture of what might happen and how to respond, then it is important to examine the visuals from the exercises, because they were literally meant to help flesh out the picture of pending devastation and response. The visuals produced from the exercise send mixed messages. While the simulation itself paints a realistic picture of the damage from a high-category hurricane, the visuals depicting the response to the simulation were not as realistic. Some of the visuals suffer from a lack of detail and urgency in their interpretation of the virtual evi-dence. Remember that the Pam exercises used virtual evidence from comput-er simulations that predicted the damage. The computer simulations them-selves merely told what would be. The planning visuals, in turn, were de-signed to suggest what course of action should follow. In terms of Lynda Wash's "is/ought" divide—or the tension between what scientific findings reveal and what we should do about it—it was the explicit and unambiguous aim of the planning visuals to map out what ought to happen in terms of rescue, recovery and relief (Walsh "Before Climategate"). Walsh describes scientific data displays that border on making recommendations. In this case, however, we have unclear and ineffective visuals that were supposed to make clear and accurate recommendations when responders needed them most.

To its credit, several elements of the workshop documents lent credibility to the event and attempted to make a compelling case to the workshop participants and planners that their preparations would suffice. For example, the sheer number of documents gave the appearance of thorough planning. Emergency response on the Gulf Coast had to incorporate several stakehold-ers, including state, local, and federal agencies, and it had to account for several moving parts in the machinery of programs and agencies responsible for emergency response. The amount of planning and explaining necessary warranted the copious number of pages. However, the numerous worksheets, workbooks, and other documents give the impression of thoroughness and attention to detail. Some features of the workbook information graphics reinforce a perception of certitude. They reinforce the perception that the coordinating organizations had the strategic plan to control unforeseen adverse events, should they occur. Some of the visuals in the documents used to communicate information also helped the documents make the case that the preparations were sufficient. For example, blueprints for flow of supplies through distribution points seem reasonable and standard for emergency events. (See figure 6.1.)

Several of the visuals show the titles of participating representatives and agencies, healthy numbers of volunteers, and other appeals to ethos and

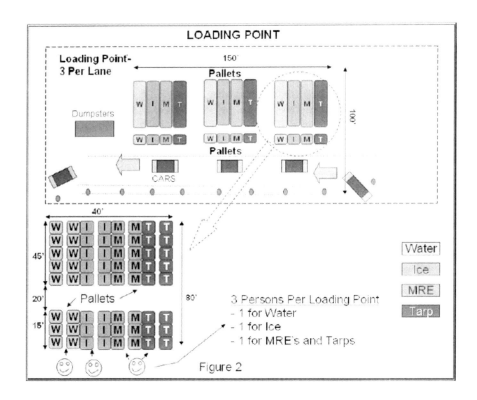

Figure 6.1.   Plan for Resource Distribution Points. Used by permission of the United States Senate Committee on Homeland Security and Governmental Affairs.

logos. However, several other of the visuals oversimplify the plans. Despite the wealth of workshop documentation and other semblances of preparation and planning, former FEMA chief Michael Brown admitted after the disaster that there was no plan for responding to the Katrina emergency (U.S. House of Representatives). Furthermore, while the Pam simulation generated a realistic picture of the threat, the Pam exercises conducted in light of the projections did not generate an equally realistic picture of all aspects of the response, rescue or recovery plan. Great tension emerged between what the simulation projected and what some of the conceptual models and flow charts outlined would be the fall-out, and rescue and recovery plan. The planning visuals take short-cuts and leave out details meant to outline the response to the simulated hurricane. The visuals would have been effective if they had matched the level of complexity and detail that the projections from Hurricane Pam predicted. Some of the visual oversimplify the scope of the response, which seems to suggest that the team did not suspend disbelief, or in this case, did not fully believe the virtual evidence of the simulation.

Take, for example, the table, map, and conceptual model from workshop binders prepared by Innovative Emergency Management Inc. (IEM), the contractors who hosted the exercises. (See figure 6.2.) In a press release about the events, they illustrate plans and highest water levels during the simulated storm. Is the meaning of simulated visuals iconographic (i.e., by resemblance) or indexical (i.e., by relationship)? Even when simulated visuals do not exact the real thing, they can still bear significant meaning. In this case, the similarity between the simulation's virtual evidence and what actually happened creates ethical tension. Visual rhetoric has a unique capacity to persuade. Simulated animations are a hybrid between photography and illustrations—they have something of the realistic value—"aura" or authenticity of photographs—without exact verisimilitude. However, this map fails to capture the fine-grained surface details of the object—that is, damaged buildings, and landmarks, cars and other recognizable objects under water. The map makes an impression because it superimposes simulated water levels (indicated by the green-colored areas) onto an actual map of Louisiana. The map does less of an effective job managing basics of visual design. For example, it does not maximize color meaning insofar as it uses green rather than blue to indicate flooding or red to denote hazardous damage.

This case broaches issues of visual ethics. If simulated visuals parallel photographs, then do they have capacity for photographic truth? In some cases, even when simulated visuals are not perfect replicas, they still manage to convey a virtual truth sufficient to teach and inform audiences. In the case of Hurricane Pam's maps and illustrations, major design elements did not paint a humane, complex, or realistic enough picture to help prepare workshop attendees. In what way should we expect verisimilitude, veracity, or virtue from simulation visuals? Sam Dragga has argued that technical visuals

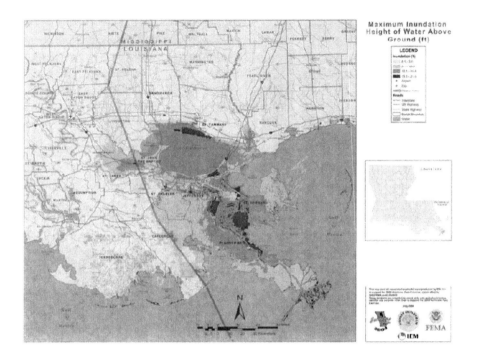

Figure 6.2.   Map of Water Levels Raised by Hurricane Pam. Used by permission of the U.S. Senate Committee on Homeland Security & Governmental Affairs.

breach ethical standards when they fail to account for the human factor behind technical information. "We must recognize the equal obligation of the visual component to support and to promote a humanized and humanizing understanding of technical subjects. In brief, ethical visuals must be has humanistic as ethical worlds" (Dragga and Voss 2001, 266). For example, this map gives us computer-precise estimates of water levels at their highest point during a Category-3 storm. This map does not, however, include population estimates at key areas of worst damage, such as Lakeview, Gentilly, and East New Orleans. Including these elements could have superimposed implications for human life and culture onto the landscape of the natural disaster. Action plans gave general directives about managing the aftermath; however, the visuals left unaddressed many detailed specifics for addressing the magnitude of the catastrophe (U.S. House of Representatives). For example, a conceptual model of the evacuation does not factor in traffic, natural obstructions, or barricades in the relief, rescue, and evacuation routes.

This case introduces some of the rhetorical and ethical dimensions that underpin the graphics and visuals designed to illustrate results from a com-

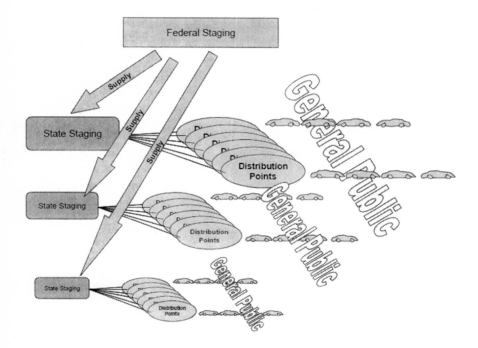

Figure 6.3.   Conceptual Model of Evacuation from Hurricane Pam. Used by permission of the United States Senate Committee on Homeland Security and Governmental Affairs.

puter simulation. It raises questions pertaining to the nature and ethics of the simulated illustration, first by engaging simulated visuals in terms of visual rhetoric, or their capacity to persuade. Computer simulations are meant to stand-in for the real thing. Simulated visuals represent virtual events in much the same way that photographs represent actual events. However, since simulated visuals represent predictions, they vary in the extent to which they capture all of the fine-grained surface details of the source—the more accurate the simulation, the more the simulated visual can have the realistic value of photographs (or what Walter Benjamin calls "aura" or authenticity). Unfortunately, in the case of visuals from Hurricane Pam, some of the visual design decisions *further removed* the representations from the potential tragedy.

Political and cultural events probably also contribute to heightened skepticism. The war in Iraq was underway. We know now that government officials mislead the public about weapons of mass destruction, and they inflated the risk and reports of chemical and biological weapons. The confluence of events might suggest a cultural and social ambivalence about facts. In the case of the Pam exercises, the simulated facts—that predicted the havoc of Katrina with some degree of accuracy—faced resistance from factions who were not convinced enough to prompt sufficient action. However, the over-statements and myths told about Iraq and the threat of Saddam Hussein were persuasive enough for the country to march to war. The visuals used to make the case for war illustrate the transference. On one hand, we see very detailed simulated facts transferred into oversimplified visuals that underplay or dis-regard those virtual facts. On the other hand, the grainy satellite images of huge bunkers in Baghdad and other circumstantial exhibits brought to the United Nations without sufficient context helped usher in a war that lasts several years and costs billions upon billions of dollars. Therefore, political priorities and regimes to truth can vest unmeritorious evidence with unwar-ranted credibility or cast doubt over the merits of ad hoc reasoning and virtual evidence.

Other studies on the gap between Pam exercises and the Katrina response point to the lack of funding for implementing recommendations post-Pam exercises and the underutilization of the plans and information from Pam in pre-landfall preparations (Congleton; Morris; Gheytanchi et al.; de Marchi; S. Williams). John Morris explains how terrorism-related exercises took precedence and delayed Pam exercises for almost three years and how FEMA had run other hurricane disaster preparation exercises prior to Pam (290). Roger Congleton posits that short term limits for local politicians facilitated turning a blind eye to the possibility that such a disaster could happen on their watch. He and Thomas Schmidlin also describe how Pam predictions led to overestimations of early casualties (Congleton 20; Schmid-lin 755). Cole and Fellows consider the Pam exercises as due diligence insofar as they served as crisis or care communication by informing stake-holders about the potential risks (216). Boin and McConnell also fault "psychological pathologies" of "overvaluation, overconfidence, insensitiv-ity" on the part of state and local authorities, "bureaucratic complexities" in executing rescue efforts, and shifted federal priorities from natural disasters to terrorism (53).

I argue that the gap between virtual evidence and virtual planning exacer-bated and underscored these problems. The Pam exercises were based on sophisticated simulation calculations of what might happen if a Category 3 hurricane hit the Gulf Coast. The visuals of the plan, however, failed to make as sophisticated an argument about rescue and recovery efforts. The work-shop's design treated as a given or a constant for the response teams who participated in the training—that the simulation's calculation would be cor-rect. However, Congressional testimony and interviews with scientists who generated the Pam simulation revealed that the emergency responders who participated in the workshop brought from their own organizational cultures and constraints that influenced how they perceived the simulated hurricane.

National politics also played a part in the praise and blame divvied out regarding the workshops. The federal government cut budgets that could have aided relief efforts. Organizational territorialism also played a role. Stakeholders from different agencies recycled standard procedures to address the simulation rather than adjusting their response to the new, simulated facts laid out before them at the workshop.

The Hurricane Pam visuals did not help communicate the event's scope to the fullest extent. Simulated visuals are hypothetical or subjunctive (i.e., indicative of what could be or might be). Similarly, ethics engage modal possibilities (i.e., what one should do). If we cannot say that subjunctive evidence is enough with which to make final decisions, then we can expect the evidence to broaden our assumptions and expectations about modal possibilities. Visuals are inherently memorable and effective for conveying meaning. Therefore, we must hold simulated visuals to high rhetorical and ethical standards, since they possess the virtue (the worth and workings)—if not the exactness—of what they represent.

DISCUSSION

Both cases demonstrate a reluctance and almost disbelief on the part of general public regarding the actionable nature of simulated facts. I believe this stems, in part, from a misunderstanding about the complexities of arguments underpinning computer simulations. The pendulum is swinging wildly between the extremes of constructivism and absolutism. To say that scientific simulations, for example, do not represent reality is not to say that they cannot replicate the virtue and practical implications of what is real, nor is it to say that all simulations are equally useful, explanatory or thorough. That several people have an argument and can make a point does not make all arguments equally valid, useful, or virtuous. Some arguments are better than others—more subtle, thoughtful, self-reflexive. This is as true for simulations as it is for expository argumentation. As Deborah Tannen argues, not all sides of an argument are created equal. Deconstruction and postmodernism have shaped how we understand arguments in general, and scientific arguments in specific. To call a simulation an argument is not to say that it, therefore, can never be representative. Something virtual is not untrue; on the contrary, consequences of virtual evidence can have the same potency and virtue as the actual evidence from the event. The virtual nature of simulations does not mean that simulated evidence has no potential. On the contrary, when simulations are done well—when they are thoughtful, self-reflexive and sophisticated—they can have the virtue of the real thing without having to be the real thing.

Both cases—Hurricane Pam and climate simulations—also speak volumes about the nature of uncertainty in science and how the public manages such uncertainty. That climate change skeptics and government officials distrusted the reliability of simulations predicting the reality of global warming or hurricane disaster projections is an extension of new debates over the handling of uncertainty in science. William Byers argues that impossible aims of science and math such as defining randomness or capturing the complexities of the natural world are important because they lead to useful scientific and technological advances, despite the fact that there "remains an inevitable gap between the definition and what is being defined" (7). Uncertainties, "incompleteness," "ambiguities," and other aspects of the development of science in the last century are the product of the advancement of rationality "filling in" the holes between what is, what we observe, and what we report about it (Byers 16). These gaps and holes, in turn, require scientists to deploy creativity in answering research questions and advancing scientific knowledge.

The tentative nature of scientific facts is essential to Thomas Kuhn's notion of scientific revolutions where huge schisms erupt and challenge the traditional paradigm, rather than advance in cohesive progression from one scientific era to another. I would add that paradigm-changing ideas persist or stand as a legitimate challenge to the tradition only when they achieve a degree of *energeia*, or potency in the middle of the action of a scientific debate. I further argue that messages increase *energeia* in the same way as did the SAS project—namely, when they do a good job of tending to audience reception and their particular and implicit uncertainties. Simulations are not an exception to the rule. In fact, they are textbook cases of science instantiating and handling uncertainty. At their very core, computer simulations require scientists to account for uncertainty, make decisions in the midst of it, and attend to disagreement, counterargument, and rebuttal with detail and care. In the case of the SAS project, the team factored counterarguments in early. They made presentations, thereby exposing their project early to audiences and their concerns, which, in turn, the SAS team internalized and redressed during composition process. In science, there have always been degrees of uncertainty, even within established science. The scientific enterprise has always been oriented toward finding facts and certainty rather than knowing it completely. Uncertainty is the reason why building consensus in science is important. When the majority of the scientific community is more certain than not, it presents the findings to the public. Facts sustain or persist only insofar as they have *energeia*, or potency in the middle of action or as they approach representing reality, rather than when they have completely represented it.

All scientific findings have degrees of uncertainty; but uncertainty is not sufficient grounds for inaction. For example, there are always degrees of uncertainty implicit in applied science. In medical science, when a randomized controlled trial finds that a treatment works for most people with a reasonably high confidence interval, the drug is approved for particular use. A statistically significant number of participants are involved in these trials, but it is impossible to know for sure whether all patients will see the same benefits. It's also impossible to know for sure whether the trials bore out all of the possible hazards or risks. Does uncertainty in such cases warrant not approving the drug? Not taking the drug? For some individual patients, one single risk would be too many to try the medication. However, when the evidence from several such trials is synthesized vis-à-vis a systematic review, and the findings more often than not point to more benefits than harms, most would agree that it would be irresponsible not to offer the medication, in spite of the risks, particularly if the benefits include decreased pain and increased quality of life. In several cases of uncertainty in medical science and public health, delayed response due to counterarguments based on one rhetorical devise—*negatio*—led to unnecessary illnesses and deaths (Michaels and Monforton). The question becomes what is an acceptable level of certainty to do something about the findings. What level of certainty warrants action? To answer the question, we should hope for arguments that expose their methods and account for complexity, strengths and weaknesses—simulations that build into their methods as many twists and turns of decision making and debate as the topic warrants—rather than the arguments that resist introspection and self-criticism.

Simulation science produces virtual, abductive evidence that factors counterargument into its reasoning. As is the case with all scientific discovery, it always involves a degree of uncertainty and, as is the case with theoretical (prospective or retrospective) science, it is speculative to a degree. What matters is whether scientists handle the speculation with a modicum of responsibility and integrity, and which of their stories accounts for several hypothetical scenarios and degrees of uncertainty. We must expect simulation science to account for the degrees of uncertainty in its own deliberations. We should expect simulation science to weigh counterarguments and strength of evidence, or lack thereof, into its analysis. It is unrealistic to expect complete certainty or the ultimate end (telos or completion) of questioning. Rather, simulations should set forth a persuasive argument that takes uncertainty and other perspectives into consideration and builds consensus. To accuse simulation science of degrees of uncertainty diminishes most of science for the same. To accuse simulation science of uncertainty also obscures the fact that skepticism proceeds from as uncertain grounds, if not more so. And uncertainty is no excuse for inaction. The best simulation science takes into account its own degrees of certainty and strength of evidence. Simply negating and disparaging simulations for not reporting phenomena with total, irrefutable accuracy holds simulated evidence to a higher standard than most of science, science that has led to helpful and significant advances in public health and general welfare. We should expect of synthesis and simulation science more than attacking a few weak points in the paradigm, particularly when the sum total of the corpus of the evidence confirms the overall argument that simulations make.

Conclusion

This book continues research on computer programming in the rhetoric of science and technical communication; in particular, it studies a much-used but often-misunderstood genre, computer simulations, and it established simulation as a rhetorical process as well as product. It examined rhetorical figures and external influences (from the scientific community and institutional affiliations) that came into play when simulation scientists eliminated factors, added artificial factors, set initial parameters, used particular algorithms, and reported their findings. It looked beyond the product and text to extra-linguistic influences meaningful to the simulation, including interactions with several audiences. Hence, this book helped uncover the relational meaning of scientific computer simulations. Furthermore, in this book, simulation was offered up as a distinct mode in the scientific imagination. Unlike the others, simulation employs abductive reasoning to produce virtual evidence, which has the potency and virtue of the real thing without being the thing itself. Simulations wield explanatory power. They help explain phenomena in ways other genres of the scientific imagination cannot. Also, rhetorical strategies helped the TSI team explain the existence of phenomenon yet to be fully explained. And the scientific community used the "news" of the SAS article to further explain and frame their own, emerging projects.

Linguistic versions of meaning could not address the multi-layered complexity of the TSI project. To understand the simulation reports and animations, the book had to move beyond merely considering psychoanalytical components of linguistic analysis; it is this psychoanalytic approach to linguistics upon which criticism from the right emerged against the climate researchers at East Anglia and that stoked the furor around Climategate. Critics took the words at face value and projected sinister motivations onto the team, rather than reading the words in their full context. This analysis can actually serve to help reveal what simulated scientific research means, given the multiple modes and venues in which professional scientists deliver their work. Applied to an entire project and its multiple documents, textual studies can retain value and avoid formal generalizations.

Finally, the book argued that simulations are abductive and virtual; in other words, they represent the virtue (or essence) of an event or phenomenon that scientists cannot easily confirm or verify with direct observation. Virtual evidence and simulations are best understood as having relational meaning; in order to understand them, we must make conspicuous the relationships between linguistic and extra-linguistic phenomena and conditions that determine their contextual value. Rather than portray a fixed and finite truth, simulations possess *energeia* (insofar as they bring before our eyes the forming of relationships between extra-linguistic and linguistic events). They reflect not only what we know, but also what we are coming to know and understand. The book has demonstrated how rhetorical strategies such as abduction, narrative, and others are instrumental in constructing, reporting and distributing the relational meaning of simulations.

We can draw several conclusions from this work. First, *rhetorical analysis can explain how simulations mean rather than simply what simulations mean.* Linguistic theories of meaning often privilege the sign and fixate upon whether or not the sign matches the truth or the reality of things. For example, the SAS simulation is virtual evidence of a distant, absent object—a supernova. So, when we ask what it means, the question implies that the referent object (e.g., the supernova, the instability) is the "what" that simulations means. In this case, the simulation does not correspond perfectly to the thing it represents. In fact, simulation scientists often knowingly impose differences, for the sake of the simulation.

The SAS analysis shows us how simulations mean (or average by bringing to bear) multiple processes, authorial investments, and social implications. Simulation code means cultural and rhetorical implications—a programmer's desire to distinguish self from other programmers, prior work on algorithms, potential contrivances caused by eliminations or impositions, and so forth. In the case of the SAS simulation, the simulation means those moments where a team sets parameters to distance or align their work with certain computational models, with organizational affiliations, and other influences. It also means not only lab work, but also e-mail catharses inspired during the recursive process of reading over drafts. The SAS simulation gained momentum between different documents regarding different activities and different discursive modes and tones. And the document trail revealed administrative, social, and interpersonal tensions in the foreground and background. Social, cultural, and institutional influences shaped the TSI team's work. The e-mails revealed how the team negotiated meanings between drafts and between each other. They also exposed how the work (and discussions about the work) changed slightly when journal reviewers weighed in and cultural demands of the astrophysical community came to bear. Once published, the work took on new significance as other researchers' cited and used it for their own ends. The SAS project also had different impact for the governmental agency for which the team worked.

Meaning is a process that produces texts; texts reflect the interconnected networks and activities that I described above. The SAS article shows that the process of meaning—gathering together and mediating different influences, ideas, and interests—leads to more meanings, which, in turn, both represent and influence action. In our lifetime, we might never know for sure if accretion shocks *actually* mechanize supernova. We can, however, see how the TSI team's ideas evolved from project to project, from their work to others' work, from e-mail drafts to journal publications. Each incarnation of the SAS story built upon prior incarnations, others' work, and social and institutional constraints. Analysis of the article and code drafts revealed how the team isolated the strongest lines of argumentation—ones that zeroed in on a consistent narrative about their research or appeased potential audience concerns and organizational demands. Analysis also revealed how their own work, passions, and suspicions about their audience compelled their decisions. Institutional affiliations, agendas, and other social influences affected how they told the story of their simulation.

What we say (and write) virtually demonstrates what we do and believe. In this way, texts are virtual evidence of meaning. The SAS article, for example, means earlier accounts of the project, institutional stakes, and input from co-authors and reviewers. Much more went into the process than what one text shows, however. For example, much more than what is reported in the final SAS article went into deciding for or against certain set parameters. Writing about it captured its essence (if not the whole thing or the real thing). In fact, it captured its potency—what people might believe about the simulation—and the audiences who read it interpreted more or less or different virtue from it.

Virtual does not mean completely without actual grounds or significance. Some work in rhetoric of science to date has positioned itself against what Dilip Gaonkar calls recalcitrance of science, or its ability to produce objective truths ("The Idea"). By drawing attention to the virtual and relational nature of simulated evidence and meaning, I do not mean to say that science has no basis in actual events, as is the case with poststructuralist notions of meaning that are preoccupied with the sign's instability. As I mentioned in chapter 2, theory mustn't confuse ideas with external realities, nor should it downplay the sign's ability to represent the world and calibrate common sense about it. Gaonkar, McGuire, Melia, Sokal, and others disagree with those who claim that language (and all its slippages) prevents science from reporting what really exists in the world. Other current philosophical and rhetorical inquiries into the conscience and self-deception also indicate that, in spite of postmodern skepticism about the sign and signified, moral and rhetorical agents still calibrate their actions and ideas by some sense of truth or reality (Budziszewski; Craig). My data does not suggest that rhetoric overturns the value of actual observation in organic chemistry or astrophysics. In fact, initial conditions of both the claisen reaction and the SAS article called upon observational data (albeit partial, at times) gathered from high-powered telescopes, microscopes, and neutrino calculators.

However, I do think it's fair to say that theoretical scientists (such as the TSI astrophysicists) are interested in what particular equations, assumptions and eliminations mean for simulated objects of study—namely, a physical phenomenon that they might not live to see with their own eyes, nor verify for themselves. In this case, they are not interested in knowledge in the traditional sense—first-hand, immediate, and so on. These scientists are interested in making sense of what they have at present—virtually materializing abstract equations and theories in the here and now—in order to continue research. In this way, facts in disciplines as obtuse as astrophysics do, in Ludwig Fleck's words, resist arbitrariness. I believe the same can be said of other theoretical sciences wherein physical verifiability takes a back seat to documenting useful meaning. Simulation scientists often don't have an observable, present object to study and know. In these cases, the simulations actualize possibilities now. They articulate a present and forecast a future. While I do not believe that this book challenges the fundamental usefulness and importance of scientific observations, it does suggest something about the production of scientific facts in theoretical sciences, in general (and astrophysics and high-end organic chemistry, in specific). In theoretical sciences, the ultimate goal might be verisimilitude, but the immediate question isn't truth so much as explanatory or narrative power. It can be argued that theoretical sciences are making theories for both the long haul and the next logical step toward better understanding.

In this way, my findings support some suppositions of realism and social constructivism and challenge others. In Bruno Latour's terms, the Janus dichotomy is not as pronounced in theoretical sciences. Latour described the tension between ready-made science and science in the making in terms of Janus, the Roman god of doorways and gates. For Latour, the Janus's two heads facing in opposite directions epitomized the tension between private science (in the throes of production) and public science (reported in journals and at conferences). One side speaks with authority and absolutism—"Science is not bent by the multitude of opinions"; and the other side probes and inquires—"How to be stronger than the multitude of opinions?" (Latour *Science*, 32). In the TSI case, the SAS article did evolve from less certain to more certain rhetorical moves. However, the Janus does not account for different ethos projected by TSI, team members who consider themselves scientists, teachers, and governmental researchers. Earlier chapters of this book illustrate how different affiliations and occasions called for different framing on the part of the TSI team. Also, the organizational story was one of perpetual promise and future development. In the annual report and the newsletters, for example, the researchers and report editors promised something newer, bigger, and better on the horizon—namely, more explicit connections between their research and energy production. Furthermore, Latour depicts a somewhat inflexible and contextually ignorant senior voice of the Janus. However, the senior researcher TSI showed clear rhetorical savvy and awareness of science in the making (during email conversations).

My findings also suggest that rhetoric does bear upon the production of orthodox, computational scientific facts. I'd argue that in simulation science, rhetorical figures and decisions do help simulation scientists produce valid claims. Take, for example, the SAS case, wherein abduction, *deliberatio*, and other rhetorical strategies enabled the team to make ad hoc decisions. However, I would agree with Alan Sokal that these rhetorical choices are part and parcel of scientific reasoning: "[T]here is no fundamental 'metaphysical' difference between the epistemology of science . . . and everyday life. . . . [A]ll human beings . . . use the same basic methods of induction, deduction and assessment of evidences as do physicists or biochemists" ("What the *Social Text* Affair Does"). Computational scientists must use sound rhetorical principles like everyone else.

Finally, relational meaning problematizes rhetorical "success." The SAS article seems successful considering the frequency with which it was cited without alteration within a few years of its publication. However, time and new research might change the tenor of the citations. The team's tendency to exclude reporting their own actions (by, for example, foregrounding the simulation and the phenomena as sentence subjects) did not prevent journal editors and reviewers from publishing the article. Which incarnation of the SAS story was more successful? The article that reviewers and editors ultimately published? The first draft to incorporate new revelations from e-mail correspondence? The earlier drafts that contained more rhetorical figures underpinning ad hoc decisions? The newsletter and report accounts that will hopefully convince the DOE to continue funding their work? There are many ways by which we might measure the SAS project a "success"—(1) when TSI receives more funding, (2) when more fellow scientists cite the work, (3) when the government puts their findings to use making alternative energies and ending the energy crisis, or (4) when others in the field confirm or extend TSI findings.

I would also hypothesize that contextual dynamics might help decide their work's rhetorical longevity. The SAS article followed on the conference presentations, articles and interviews where the TSI team trumpeted their work. In fact, the team emailed messages to one another about the urgency to publish the work because the scientific community was already abuzz with a version of the news. The article's continued success might also partially depend on further, similar promotion. The published work, then, only communicates part of the paper's success. Indeed, depending on how you define rhetorical success, other incarnations of their report might matter more.

This book has implications for generic analysis of computer simulation articles. It would be interesting to see if computational science is evolving a different kind of scientific article. Gross and Harmon identify five kinds of scientific reports with different aims. Experimental articles recount lab work. Observational articles describe natural objects in nature. Mathematical articles explain causes for physical relationships. Theoretical articles offer concepts that might confirm those explanations. Methodological articles present means for doing experiments and running tests. If the SAS article represents another genre, then it appears to combine features from all of the above, insofar as simulations incorporate lab work (tweaking code and incorporating findings from nuclear physics), mathematical calculations and translations, and theoretical and mathematical implications.

According to my findings, computational articles might differ in organization, citations, hedging, and other features. Compared to the typical modern research article (as described in Gross, Harmon, and Reidy), the SAS article contained separate methods sections for computational and mathematical equations and fewer hedges overall. The SAS article demonstrates the forward tone that scientists often use when making claims about their work and refuting the alternatives—the paper tended to emphasize the phenomena over their methods. And, ultimately, despite the groundbreaking news, the SAS introductions also tended to frame the news in terms of continuing research rather than filling a gap in the field. Perhaps other computational articles share these and other characteristic features. Future studies could help catalog the features of computational articles.

This book also has implications for the study of the epistemology of science. My work complements studies of the epistemology of science by helping show the precise mechanisms and rhetorical strategies through which scientific pursuits such as computer simulations retain creative and social components. Typically critics trace the origins of modern science to Western influences and the European Enlightenment when academic disciplines such as biology, mathematics, astronomy, physics, chemistry, and medicine began reifying procedures of investigation into natural phenomenon and dispelling commonly held beliefs about the way the world worked. From Descartes, Newton, and other forebears, scientific endeavors inherited a deep reliance on rationalism (which anchors knowledge in reasoning) and empiricism (which anchors knowledge in sensory experience). Middle Eastern scholars such as Alhazen also developed a system of observing the world, noticing a problem, making a prediction about the explanation or solution to the problem, testing the prediction, and understanding and sharing the results—often considered the origins of the scientific method. This method of systematic inquiry predated the European Enlightenment by hundreds of years, but it could be argued that it, too, heavily depended on observation and reason or drawing conclusions.

Usually, historians then note a critical turn in the mid-1900s when Einstein's thought experiments and unorthodox, but inspired insights on physics yielded theories that help the scientific community better understand the phenomena of time, space, light, and motion. Turning from a heavy reliance on observation—for example, some of the phenomena he described he had not actually seen with his naked eye—Einstein explained how faith was at the foundation of rational, scientific thought: "[T]he belief in the existence of basic all-embracing laws in Nature also rests on a sort of faith" (Einstein *Albert Einstein: The Human Side*, 32). Theories of scientific knowledge have evolved from positivism (or looking at the scientific endeavor as a process of accumulating observations and verifying their validity) toward perspectives closer to constructivism (or seeing science as influenced by human and social values and experiences).

Other revisionary perspectives on scientific discovery have also emerged. Karl Popper described how scientific fact is not established positively, but rather negatively. According to Popper, the scientific community works to falsify rather than confirm theories. Then Thomas Kuhn explained how revolutions propel scientific change. According to Kuhn, science advances when radicals challenge the status quo, rather than when facts line up and accumulate. When one paradigm of science shifts, when it replaces another, the tools of the older paradigm cannot serve to check or disprove the newer one. Other theories dispel the misconception that accumulating any amount of evidence can ever completely prove any scientific theory indefinitely (Horner and Rubba; Lopushinsky). Newer epistemologies of science acknowledge the relationship between subjectivity and observation, assumptions and reason. They deem a priori justification and social dynamics essential for scientific knowledge. By examining the rhetorical nature of simulations used for the purposes of scientific discovery, this book catalogs the anatomy of both the convictions and social expectations that help scientists make decisions about the parameters of simulations. Scientists use rhetorical strategies to address audience concerns and predict and address particular questions, expectations, and reservations about the subject matter. They also use those figures to organize and justify the decisions and concessions they make regarding the subject matter.

Finally, this book also has implications for the gaming studies. The argument that this book presents contributes to the conversation currently underway in digital humanities and game studies regarding procedural rhetoric. Game studies examine games in culture in terms of the games themselves, the players who use them, and the context created with the interaction of games and players. Game studies and simulation studies have roots in similar histories. Both owe a great debt to Vannevar Bush, Ted Nelson, Espen Aarseth, Janet Murray, Jay David Bolton, and other scholar-forebears in the study of hypertexts—or electronic text that link and play multimedia and that allow users to interact in ways that extend simple reading. Both simulations and video games can employ heavy mediation—sound, video, color, image, and so forth—and both are not static texts (i.e., pages upon which words are printed for people to read), but rather layered and kinetic insofar as they are constructed of moving images or sub-routines that produce several iterations of data to analyze or interpret. Game studies traditionally revolve around questions regarding whether games have more potential for their storytelling ability or for their interactive capacity. Both perspectives hold that games derive meaning from the distance or proximity of action in games to our socially and culturally constructed perceptions of reality. Therefore, games can tell stories about capitalism, or belonging, or other humanistic values, and we can learn from them in this regard. Or game creators can build into the worlds that they create socially and culturally constructed assumptions about how politics, identity, and other humanistic systems and traits manifest. What games mean derives from the act of play itself. As David Myers puts it, computer games have an "anti-aesthetic" insofar as the significance of video games is achieved in the act of playing video game itself, and, in the act of play, a "form of real-world experience" is created for each game player. Furthermore, Roger Caillois explains that "simulation games" are one of four game types (including competition games, games of chance, and physical games) that involve "mimicry" of the real world. This aspect of game studies resembles the argument made in this book about virtual evidence and its capacity for containing the virtue and potency of the real thing itself. In this case, the act of playing a game, according to many game studies theorists, contains the virtue of a lived experience.

Furthermore, this book has something to offer game theorists who analyze the core (or the procedural nature of game play) of games (rather than their shell, or symbolic representation through moving images, perspective, etc.). The game core is the set of rules (in a board game, for example) and algorithms (in a video game) that constrain the boundaries of the action of the game. Ian Bogost offers a very compelling argument about the nature in the game core. He developed the concept of "procedural rhetoric," or "using processes persuasively . . . [and] authoring arguments through processes" (Bogost 28–29]. He explains that the rules of video games are written in the super- and sub-routines of the code undergirding the game. For many interactive and free-form games, it also includes the rules and procedures that players impose on the game or are permitted to set by the game itself through conversations among players. According to Bogost, when a game presents a rule where the avatar must make money because he or she wants to buy expensive things, it constructs and proselytizes a particular ideology, which Bogost describes in terms of enthymemes that describe the world. In this case the game posits that expensive things are good. It is the unstated assumption or what Bogost calls the "simulation gap" that you have to believe if you buy into the proposition that the avatar should make money because expensive things cost money, and the avatar needs expensive things. In setting up this simulation gap, the game posits a very particular ideology with regards to capitalism and consumerism. Bogost helps articulate the logic of play from the perspective of the player, and he intimates the kind of logic that game developers might bring to bear when scripting the various scenarios of action of the game. This book expands categories of premises and claims that developers might employ in creating the "simulation gap" of which Bogost speaks. Future studies might examine transcripts from story meetings, storyboard drafts, and other preliminary planning and development tools used by game designers to pinpoint the specific rhetorical devices they use to draw conclusions and make assumptions about how the world works and how to represent it in the game's core.

While the rhetorical strategies that underpin computer simulations and ad hoc decision making in science might very well be similar to those that underpin the decision making that game creators use to orchestrate the core of a video game, there are differences between scientific simulations and games. The stakes of video games are different than for scientific simulations. For the former, the primary objective is either fun (entertainment) or education (pedantic). In science, the primary purposes are exploration and discovery. Ultimately, scientific simulations set out to answers to questions for the purposes of advancing scientific discovery and knowledge. In turn, these findings have political and social implications insofar as policy makers interpret and operationalize their significance (or fail to do so at their peril or to their political advantage). Games needn't have any of these dimensions to qualify as a simulated game. Furthermore, science simulations aren't always interactive. Games are inherently interactive; even in instances of one-player games (such as computer versions of solitaire), the objective for game players is to beat their old score or performance and, thus, beat themselves as a prior player. The point of science simulations is not necessary immersion—at whatever level of user experience, be it using vivid 3D graphics, or a visor that alters the visually perceived world, or a chamber in which an altered reality is presented—but rather observation and deliberation. Scientific simulations can enlist multimodal observations by incorporating multimedia in the rendering of findings. In this regard it can be immersive, but the point of immersion would be for observational purposes.

Assessing the efficacy of games is difficult, particularly if the point of games is pleasure or education. First, pleasure is an aesthetic value, which can be indirectly useful for social or cultural amelioration, but not as directly useful as in the case of data produced by scientific simulations. Second, while some studies suggest the value of learning in multimodal and simulated environments, the jury is still out on whether gaming has practical implications. Even if gaming produces a virtual mode of existence or experience from which people are capable of learning social rules, language, identity, and self-awareness, the link between what is learned in abstract and what is applied in truth is still unclear. The practical implications of gaming have roots in developmental psychology, child psychology, and the role of play in learning social rules, language, identity, and self-awareness. However, it is unclear the extent of the usefulness of games for adult learning and social or cultural change. For example, several studies dispel the notion that violent video games cause or compel violent behavior in children. On the other hand, studies are still determining the extent to which gaming can change deeply and long-held negative health behaviors. On the contrary, in the case of the SAS article, the usefulness or explanatory power of scientific simulations is often immediately practical for the advancement of scientific knowledge.

FUTURE DIRECTIONS

Future work could translate lessons learned from computational simulations to critique using simulation for teaching purposes. Simulations, service learning, and other attempts to bring the "real world" into the classroom can smack of pretense and feel very contrived to student participants. Classroom simulations often fail because (1) they neglect to identify and isolate the essential differences between the workplace activity they mean to simulate and (2) they do not fully factor in those differences when making eliminations, additions, initial parameters, and other ad hoc decisions necessary for building simulations. For example, to help achieve a semblance of workplace essence, an instructor might construct a memo assignment by which students must relay data to one another for completing the next major assignment or another course in the degree plan. In this way, the memo serves another end—namely, helping finish another assignment. In order to retain the virtue of the experience writing in workplace settings, instructors will have to make some ad hoc eliminations and substitutions to address the differences between workplace and classroom writing. For example, rather than asking students to write for external clients with whom students will have no lingering contact, instructors might choose a client within the university setting which students can immediately identify and in which students have an immediate stake (e.g., writing to help the department or university accomplish some goal).

Future empirical studies could use surveys to assess whether the rhetorical moves that the TSI team used are true of most other simulations. It would be interesting to see if other simulation scientists use the same rhetorical figures at the same integral moments. For example, TSI used *pareuresis* when justifying how they set initial conditions of their simulation—do other teams use the same rhetorical moves to set similar parameters? Also, longitudinal studies could assess how time, audience needs, and other pressures in the scientific community altered the news regarding the SAS project. Ethnographic observations (visits to computational science labs themselves) could help trace institutional habits that further shape and constrain simulation science—I only studied one of multiple teams funded by the DOE. More points of comparison—for example, studying more DOE research teams—could help identify whether DOE objectives expand or constrain the teams' rhetorical productivity. All of these would aid the self-reflexivity that computational science now undergoes.

Future work should also further investigate the visual components of the simulation product. Gross and Harmon found that scientific articles in the twentieth century also include graphic representations of information; 100 percent in the final quarter of the century contained numbered figures with titles; 47 percent had numbered tables with titles; and 40 percent had numbered equations (Gross and Harmon 173). Future audience analyses of simulated animations are in order. Take, for example, how the TSI team used color-coded animations to illustrate the surges and waves emanating from the core of their model. Unlike illustrations and photography, these animations move through time. In this way, they resemble the supernova themselves more so than illustrations. However, unlike digital or analog recordings, these animations do not capture fine-grained surface details of the object; simulated animations often require a key to help readers understand what the colors and motions signify. Simulated animations are a hybrid between photography and illustrations—they have something of the realistic value of photographs, without the image verisimilitude. The visual products of simulations also raise questions pertinent to visual ethics and photographic truth. TSI used stills from animations to illustrate changes in concentric waves that evidence the instability. Jane Wang used similar animations of bumblebee wing motion. When we look at the stills, we aren't viewing real accretion shocks or bee wings, but something about having the illustrations makes the findings seem more real. Do simulated illustrations and animations breach standards of ethics—for example, Joel Feinberg's notions of prima facie obligations (in particular veracity or truth-telling and fidelity or doing what you say or promise)? Concepts from rhetorical analysis, visual rhetoric and usability studies would serve to aid computational scientists in this endeavor.

Future projects might also include a simulation programming stylebook. If it is true that rhetoric factors prominently in how simulation scientists build, frame, and report their work, then rhetoric might help simulation scientists improve software and simulation development. So far, they use syntactic and semantic validation procedures applied to any other kind of computer code. However, reading and verifying simulations might also require assessing citations, unstated assumptions implicit in equations used and eliminations or additions made, and so on. No simulation textbooks address ad hoc decision making or the rhetorical strategies that underpin them. The lack of rhetorical content in simulation textbooks might also explain why many introductory and advanced sections of simulation courses do not contain units wherein the class discusses ad hoc or rhetorical skills necessary for building simulations (Shiflet; Fell, Proulx, and Casey; Schneider, Schwalbe, and Halverson).

Some have found that rhetoric improves how instructors teach computer programming. Joanna Wolfe, for example, surveyed eighty-one computer science students about different computer science instructions. She found that those students preferred assignments that included information about real-world contexts and people. "[B]y making small shifts in the wording of their assignments, CS instructors may be able to improve student satisfaction with their classes as well as better prepare students for the human-centered demands of the contemporary workplace" (158–159). Fledgling computer programmers need rhetorical savvy to comprehend and successfully complete class assignments. I argue that the necessity of rhetorical know-how does not end for college-level computer programmers, nor does it solely train them for dealing with actual clients. It does not merely make assignments more contextualized, interesting, and engaging. Rhetorical strategies help rationalize the ad hoc reasoning necessary for creating simulation programs.

Bibliography

Achinstein, Peter. *The Book of Evidence*. New York: Oxford University Press, 2001. Print

Advanced Scientific Computing Advisory Committee (ASCAC). *Minutes for the Advanced Scientific Computing Advisory Committee Meeting. 13 March 2003*. 2003. PDF file.

Anderegg, William R.L., James W. Prell, Jacob Harold, and Stephen H. Schneider. "Expert Credibility in Climate Change." *Proceedings of the National Academy of Sciences of the United States of America*. 107.27 (2010): n. pag. Web. 7 June, 2013. http://www.pnas.org/content/107/27/12107.long

Arbib, Michael A. *The Construction of Reality*. Cambridge: Cambridge University Press, 1986. Print.

Aristotle. *The Basic Works of Aristotle*. Trans. Richard McKeon. New York: Random House, 1941. Print.

Baake, Ken. *Metaphor and Knowledge: The Challenges of Writing Science*. Albany: State University of New York Press, 2003. Print.

Bacon, Francis. *The Advancement of Learning*. Oxford: Clarendon Press, 1926. Print.

Balantekin, A. B. and G. M. Fuller. "Supernova Neutrino–Nucleus Astrophysics." *Journal of Physics G: Nuclear and Particle Physics* 29.11 (2003): 2513. PDF.

Barrow, John. "Living in a Simulated Universe." *Universe or Multiverse*. Ed. Bernard Carr. Cambridge: Cambridge University Press, 2007. 481–486. PDF file.

Baudrillard, Jean. *Simulacra and Simulation*. Trans. S. Glaser. Ann Arbor: University of Michigan Press, 1994. Print.

Bazerman, Charles. *Constructing Experience*. Carbondale: Southern Illinois University Press, 1994. Print.

———. *Shaping Written Knowledge: The Genre and Activity of the Experimental Article in Science*. Madison: University of Wisconsin Press, 1988. Print.

Beere, Jonathan. *Doing and Being: An Interpretation of Aristotle's Metaphysics Theta*. New York: Oxford University Press. 2012. Print.

Benjamin, Walter. "The Work of Art in the Age of Mechanical Reproduction." *Illuminations*. Trans. Harry Zohn. New York: Schocken Books, 1969. 217–251. Print.

Berkenkotter, Carol and Thomas N. Huckin. "You Are What You Cite: Novelty and Intertextuality in a Biologist's Experimental Article." *Genre Knowledge in Disciplinary Communication*. Hillsdale, NJ: Lawrence Erlbaum Associates, 1995. 45–61. Print

———. *Genre Knowledge in Disciplinary Communication: Cognition/Culture/Power*. Mahwah: Lawrence Erlbaum Associates, 1995. Print.

Berlin, James. "Contemporary Composition: The Major Pedagogical Theories." *College English* 44 (1982): 765–777. Print.

————. *Writing Instruction in Nineteenth-Century American Colleges: Studies in Writing and Rhetoric.* Carbondale: Southern Illinois University Press, 1984. Print.

Bernardini, Silvia. "Using Think-Aloud Protocols to Investigate the Translation Process: Methodological Aspects." *RCEAL Working Papers in English and Applied Linguistics 6.* Ed. John N. Williams. Cambridge: University of Cambridge Press, 1999. 179–199. PDF file.

Biesecker, Barbara A. "Rethinking the Rhetorical Situation from within the Thematic of Différance." *Philosophy and Rhetoric* 22.2 (1989): 110–130. 15 March 2006. Print.

Bitzer, Lloyd. "The Rhetorical Situation." *Philosophy and Rhetoric* 1 (1968): 1–15. Print.

Black, Max. *Models and Metaphors: Studies in Language and Philosophy.* Ithaca: Cornell University Press, 1962. Print.

Blackie, D. "Science and Socialists: It's All Relative." *Socialist Worker Review* 115 (1988): 26–27. Web. 12 May 2013. http://www.marxisme.dk/arkiv/blackied/1988/einstein.htm.

Blondin, John. "VH-1." September 1, 1990. Computer code. 10 May 2012. Web. http://wonka.physics.ncsu.edu/pub/VH-1.

Blondin, John, Anthony Mezzacappa, and Christine DeMarino. "Stability of Standing Accretion Shocks, with an Eye Toward Core-Collapse Supernova." *Astrophysical Journal* 30. 584 (2003): 971–980. Print.

Blondin, John and Anthony Mezzacappa. "The Inherent Asymmetry of Core-Collapse Supernovae. American Astronomical Society, HEAD Meeting #7, #10.13." *Bulletin of the American Astronomical Society* 35 (2003): 613. Web.12 May 2013. http://adsabs.harvard.edu/abs/2003HEAD....7.1013B.

————. "Sources of Turbulence in Core Collapse Supernovae. American Astronomical Society, 199th AAS Meeting, #94.06." *Bulletin of the American Astronomical Society* 33 (2001): 1445. Web. 9 May 2013. http://adsabs.harvard.edu/abs/2001AAS...199.9406B.

Bloor, David. "The Strengths of the Strong Programme." *Scientific Rationality: The Sociological Turn.* Dordrecht, MA: D. Reidel Publ. Company, 1984. Print.

Bogost, Ian. *Persuasive Games: The Expressive Power of Videogames.* Cambridge: MIT Press, 2007. Print.

Boin, Arjen and Allan McConnell. "Preparing for Critical Infrastructure Breakdowns: The Limits of Crisis Management and the Need for Resilience." *Journal of Contingencies and Crisis Management* 15.1 (2007): 50–59. Print

Bokeno, R. Michael. "The Rhetorical Understanding of Science: An Explication and Critical Commentary." *The Southern Speech Communications Journal* 52 (1987): 28–311. Print.

Bokulich, Alisa. "Rethinking Thought Experiments." *Perspectives on Science* 9.3 (2001): 285–30. Print.

Bolter, Jay David and Richard Grusin. *Remediation: Understanding New Media.* Cambridge: MIT Press, 1999. Print.

Bormann, Ernest G. "Fantasy and Rhetorical Vision: The Rhetorical Criticism of Social Reality." *Quarterly Journal of Speech* 58 (1972): 396–407. Print.

————. "Fantasy and Rhetorical Vision: Ten Years Later." *Quarterly Journal of Speech* 68 (1982): 289. Print.

Bostrom, Nick. "Are You Living in a Computer Simulation?" *Philosophical Quarterly* 5.211 (2003): 243–255. Print.

Brooks, Michael. "Life's a Sim and Then You're Deleted." *New Scientist* 175.2353 (2002): 48–50. Print.

Brown, Harold I. *Rationality.* New York: Routledge, 1988. Print.

Brown, J. A. C., H. S. Houthakker, and S. J. Prais. "Electronic Computation in Economic Statistics." *Journal of the American Statistical Association* 48.263 (1953): 414–428. Print.

Brown, James Robert. *The Laboratory of the Mind: Thought Experiments in the Natural Sciences.* New York: Routledge, 1991. Print.

Brown, Theodore L. *Making Truth: Metaphor in Science.* Champaign: University of Illinois Press, 2003. Print.

Budziszewski, J. "Handling Issues of Conscience in the Academy." *The Newman Rambler* 3. 2 (1999): 2–9. Web. 2 June 2012. http://catholiceducation.org/articles/religion/re0289.html.

Byers, William. *The Blind Spot: Science and the Crisis of Uncertainty.* Princeton: Princeton University Press. 2011. Print.

Caillois, Roger. *Man, Play, and Games*. Champaign: University of Illinois Press, 2001. Print.

Campbell, John A. "Darwin, Thales, and the Milkmaid: Scientific Revolution and Argument from Common Belief to Common Sense." *Perspectives on Argument*. Eds. Robert Trapp and Janice Schuetz. Prospect Heights, IL: Waveland, 1990. 207–220. Print.

Casti, John L. *Would-Be Worlds: How Simulation Is Changing the Frontiers of Science*. New York: J. Wiley, 1997. Print.

Ceccarelli, L. *Shaping Science with Rhetoric: The Cases of Dobzhansky, Schrodinger, and Wilson*. Chicago: The University of Chicago Press, 2001. Print.

———. "Manufactured Scientific Controversy: Science, Rhetoric, and Public Debate." *Rhetoric & Public Affairs* 14.2 (2011): 195–228. Print.

Charney, Davida. "From Logocentrism to Ethocentrism: Historicizing Critiques of Writing Research." *Technical Communication Quarterly* 7.1 (1998): 9–32. Print.

Cherwitz, Richard and James Hikins. *Communication and Knowledge: An Investigation in Rhetorical Epistemology*. Columbia, South Carolina: University of South Carolina Press, 1986. Print.

Cherwitz, Richard A. and James W. Hikins. "Irreducible Dualism and the Residue of Common-sense: On the Inevitability of Cartesian Anxiety." *Philosophy and Rhetoric* 23 (1990): 229–241. Print.

Cicero. *On Oratory and Orators*. Trans. J. S. Watson. Carbondale: Southern Illinois University Press, 1986.

Citizens for Responsibility and Ethics in Washington (CREW). *The Best Laid Plans: The Story of How the Government Ignored Its Own Gulf Coast Hurricane Plan*. 27 June 2007. Web. 30 May 2013. http://www.citizensforethics.org/page/-/PDFs/Reports/Katrina%20DHS%20Report.pdf?nocdn=1.

Cochrane Collaboration. "Levels of Evidence." *Cochrane Consumer Network (CCNET)*. Cochrane Collaboration, 25 June 2012. Web. 18 June 2013. http://consumers.cochrane.org/levels-evidence.

Cole, Terry W. and Kelli L. Fellows. "Risk Communicaiton Failure: A Case Study of New Orleans and Hurricane Katrina." *Southern Communication Journal* 73.3 (2008): 211–228. Print.

Colella, Phillip, et al. "A Science-Based Case for Large-Scale Simulation." Washington, D.C.: DOE Office of Science (2003). PDF.

Coleridge, Samuel Taylor. *Biographia Literaria*. Project Gutenberg: 2004. Web. 15 May 2013. http://www.gutenberg.org/files/6081/6081-h/6081-h.htm.

Congleton, Roger D. "The Story of Katrina: New Orleans and the Political Economy of Catastrophe." *Public Choice* 127 (2006): 5–30. Print.

Craig, William Lane. "Robert Adam's New Anti-Molinist Argument." *Philosophy and Phenomenological Research* 54.4 (1994): 857–861. Print.

de Marchi, Bruna. "Not Just a Matter of Knowledge: The Katrina Debacle." *Environmental Hazards* 7.2 (2007): 141–149. Print

Dear, Peter. *The Literary Structure of Scientific Argument: Historical Studies*. Philadelphia: University of Pennsylvania Press, 1991. Print

———. *Discipline and Experience: The Mathematical Way in the Scientific Revolution (Science and Its Conceptual Foundations)*. Chicago: University of Chicago Press, 1995. Print.

Di Paolo, Ezequiel A., Jason Noble, and Seth Bullock. "Simulation Models as Opaque Thought Experiments." *Artificial Life VII: The Seventh International Conference on the Simulation and Synthesis of Living Systems*. Portland, Reed College, 2000. Web. 12 May 2013. http://www.sussex.ac.uk/Users/ezequiel/opaque.pdf.

Dietrich, Frank. "The Computer: A Tool for Thought-Experiments." *Leonardo* 20.4 (1987): 315–325. Print.

Einstein, Albert. "Autobiographical Notes." *Albert Einstein: Philosopher-Scientist*. Ed. A. Schlipp, La Salle: Open Court, 1949: 53. Print.

———. *Albert Einstein: The Human Side*. Eds. Helen Dukas and Banesh Hoffman. Princeton: Princeton University Press, 1979. Print.

Eisenberg, Anne. "Metaphor in the Language of Science." *Scientific American* May 1992: 144. Print.

Elshoff, James and Michael Macrotty. "Improving Computer Program Readability to Aid Modification." *Communications of the ACM* 25.8 (1982): 512–521. Print.

Fahnestock, Jeanne. *Rhetorical Figures in Science.* New York: Oxford UP. 1999. Print.

Fahnestock, Jeanne and Marie Secor. "The Stases in Scientific and Literary Argument." *Written Communication* 5.4 (1988): 427–443. Print.

Faigley, Lester. "Nonacademic Writing: The Social Perspective." *Writing in Nonacademic Settings.* Eds. Lee Odell and Dixie Goswami. New York: Guilford. 1985. 231–248. Print.

Feinberg, Joel. "Civil Disobedience in the Modern World." *Humanities in Science* 2.1 (1979): 37–60. Print

Fell, Harriet, Viera K. Proulx, and John Casey. "Writing Across the Computer Science Curriculum." *SIGCSE '96 Proceedings of the Twenty-Seventh SIGCSE Technical Symposium on Computer Science Education* 28.1 (1996): 204–209. Web. 7 April 2013. http://dl.acm.org/citation.cfm?id=236540.

Flanagan K. A., C. R. Canizarez, D. Dewey, J. C. Houck, A. C. Fredericks, M. L. Schattenburg. T. H. Markert, and D. S. Davis. "CHANDRA High-Resolution X-Ray Spectrum of Supernova Remnant." *Astrophysical Journal* 605 (2004): 230–246. Print.

Fleck, Ludwig. *Genesis and Development of a Scientific Fact.* Eds. Thaddeus Trenn and Robert Merton. Trans. Frederick Bradley. Chicago: University of Chicago Press: 1981. Print.

Foley, Henry C., Alan W. Scaroni, and Candace A. Yekel. "RA-10 Inquiry Report: Concerning the Allegations of Research Misconduct Against Dr. Michael E. Mann. Department of Meteorology, College of Earth and Mineral Sciences." Pennsylvania State University. 2010. Web. 14 May 2010. http://www.research.psu.edu/orp/documents/Findings_Mann_Inquiry.pdf.

Galloway, Alexander. *Protocol: How Control Exists after Decentralization.* Cambridge: MIT Press, 2004. Print.

Gaonkar, Dilip P. "The Idea of Rhetoric in the Rhetoric of Science." *Rhetorical Hermeneutics: Invention and Interpretation in the Age of Science.* Eds. Alan G. Gross and William Keith. Albany: State University of New York Press. 1997. 25–89. Print.

Gheytanchi, Anahita, et al. "The Dirty Dozen: Twelve Failures of the Hurricane Katrina Response and How Psychology Can Help." *American Psychologist* 62.2 (2007): 118–130. Print.

Gould, Harvey, Jan Tobochnik, and Wolfgang Christian. *An Introduction to Computer Simulation Methods: Applications to Physical Systems.* Reading: Addison-Wesley. 1996. Print.

Gragson, Gay and Jack Selzer. "Fictionalizing the Readers of Scholarly Articles in Biology." *Written Communication* 7 (1990): 25–58. Print.

Grice, Paul. *Studies in the Way of Words.* Cambridge: Harvard University Press. 1989. Print.

Gross, Alan. *The Rhetoric of Science.* Cambridge: Harvard University Press. 1996. Print.

Gross, Alan, Joseph Harmon and Michael Reidy. *Communicating Science: The Scientific Article from the 17th Century to the Present.* New York: Oxford University Press. 2002. Print.

Gunn, Joshua. "Refiguring Fantasy: Imagination and Its Decline in U. S. Rhetorical Studies." *Quarterly Journal of Speech* 89. 1 (2003): 41–59. Print.

Hanson, Robin. "How to Live in a Simulation." *Journal of Evolution and Technology* 7 (2002). Web. 4 April 2013. http://www.jetpress.org/volume7/simulation.htm.

Harris, Paul. "Thinking About What Is Not The Case." *International Journal of Psychology* 28.5 (1993): 693–707. Print.

Harris, Randy Allen. "Reception Studies in the Rhetoric of Science." *Technical Communication Quarterly* 14.3 (2005): 249–255. Print.

Herant, Mark, Willy Benz, W. Raphael Hix, Chris L. Fryer, and Stirling A.Colgate. "Inside the Supernova: A Powerful Convective Engine." *Astrophysical Journal* 435.1 (1994): 339–361. Print

Herant, Marc, Stirling A. Colgate, Willy Benz, and Chris Fryer. "Neutrinos and Supernovae." *Los Alamos Science* 25 (1997): 64–79. Print.

Hoffman, Michael. "Problems with Peirce's Concept of Abduction." *Foundations of Science* 4.3 (1999): 271–305. Print.

Hoffman, Andrew J. "Talking Past Each Other? Cultural Framing of Skeptical and Convinced Logics in the Climate Change Debate." *Organization Environment* 24.1 (2011): 3–33. Print.

Holton, Gerald. "On the Art of the Scientific Imagination." *Daedalus* 125.2 (1996): 186–209. Print.

———. *Scientific Imagination.* Cambridge: Harvard University Press, 1998. Print.

Horner, J. and Peter Rubba. "The Laws are Mature Theories Fable." *The Science Teacher* 46.2 (1979): 31. Print.

Hughes-Etzkorn, Letha. "A Metrics-Based Approach to the Object-Oriented Reusable Software Components." Diss. University of Alabama, 1997. Ann Arbor: UMI, 1997. PDF file.

Hulme, Mike. "Mediated Messages about Climate Change: Reporting the IPCC Fourth Assessment in the UK Print Media." *Media and Climate Change.* Eds. In T. Boyce and J. Lewis. New York: Peter Lang, 2009. Print

HyperChem. Hypercube, Inc. 5 March 2003. Web. 3 April 2013. http://www.hyper.com.

Intergovernmental Panel on Climate Change (IPCC). *Contribution of Working Group I to the Fourth Assessment Report of the Intergovernmental Panel on Climate Change, 2007.* Cambridge, NY: Cambridge University Press, 2007. Web. 5 June 2012. http://www.ipcc.ch/publications_and_data/ar4/wg1/en/contents.html.

———. *Contribution of Working Group III to the Fourth Assessment Report of the Intergovernmental Panel on Climate Change, 2007.* Cambridge, NY: Cambridge University Press, 2007. Web. 5 June 2012. http://www.ipcc.ch/publications_and_data/ar4/wg3/en/contents.html.

Irvine, A. D. "Russell's Paradox." *Stanford Encyclopedia of Philosophy.* Ed. Edward N. Zalta. Spring 2009. Web. 9 May 2013. http://plato.stanford.edu/entries/russell-paradox/.

Iverson, Brent. *Iverson Lab.* 2012. Web. 13 March 13. http://iverson.cm.utexas.edu/research/Iverson2013/Iverson_Lab.html.

Jackson, Stanley. "The Imagination and Psychological Healing." *Journal of the History of the Behavioral Sciences* 26 (1990): 345–358. Print.

Jamieson, Kathleen. "Generic Constraints and the Rhetorical Situation." *Philosophy and Rhetoric* 7 (1974): 162–170. Print.

Jung, Carl. *Memories, Dreams, Reflections.* Ed. Aniela Jaffe. Trans. Richard and Clara Winston. New York: Pantheon, 1961. Print.

Kernighan, Brian W. and Rob Pike. *The Practice of Programming.* Reading: Addison-Wesley, 1999. Print.

Kernighan, Brian W. and P.J. Plauger. *The Elements of Programming Style.* Columbus: Mcgraw-Hill, 1978. Print

Keyes, David. *Building a Science-Based Case for Large-Scale Simulation.* SCALES. New York: Columbia University. 2003. Web. 2 May 2013. http://www.cs.odu.edu/~keyes/scales/news_release.doc.

Kinneavy, James. *A Theory of Discourse.* Englewood Cliffs: Prentice Hall, 1971. Print.

Kövecses, Zoltan. *Metaphor: A Practical Introduction.* New York: Oxford University Press, 2002. Print.

Krell Institute. "A Scientific Star." *DEIXIS: The DOE CSGF Magazine.* 2004. Web. 3 May 2013. http://www.krellinst.org/doecsgf/deixis/2004/research.php?id=321.

Kuhn, Thomas S. *The Structure of Scientific Revolutions.* Chicago: University of Chicago Press, 2012. Print.

Kupferschmid, Michael. *Classical FORTRAN: Programming for Engineering and Scientific Applications.* New York, NY, USA: Marcel Dekker Incorporated, 2002. Print.

Lanham, Richard A. *A Handlist of Rhetorical Terms.* Berkeley: University of California Press, 1991. Print.

Latour, Bruno. *Pandora's Hope: Essays on the Reality of Science Studies.* Cambridge: Harvard University Press, 1999. Print.

———. *Science In Action: How to Follow Scientists and Engineers through Society.* Cambridge: Harvard University Press, 1987. Print.

———. *We Have Never Been Modern.* Cambridge: Harvard University Press, 2012. Print.

Latour, Bruno and Steve Woolgar. *Laboratory Life: The Social Construction of Scientific Facts.* Cambridge: Princeton University Press, 1979. Print.

LeFevre, Karen B. *Invention as a Social Act.* Carbondale: Southern Illinois University Press, 1987. Print.

Levinson, Paul. *Realspace: The Fate of Physical Presence in the Digital Age, On and Off Planet.* New York: Routledge, 2003. Print.

Lopushinsky, T. "Does Science Deal in Truth?" *The Journal of College Science Teaching* 23 (1993): 208. Print.

McGuire, J. E. and Trevor Melia. "Some Cautionary Strictures on the Writing of the Rhetoric of Science." *Rhetorica* 7 (1989): 87–99. Print.

McMaster, Joe, Peter Standring, and Tom Stubberfield. "The Man Who Predicted Katrina." *NOVA.* PBS, 22 Nov. 2005. Web. 8 June 2013. http://www.pbs.org/wgbh/nova/earth/predicting-katrina.html.

Mezzacappa, Anthony. "Predicting the Gravitational Wave Signatures of Core Collapse Supernovae: The Road Ahead." *Presentation at the Gravitational Wave Source Simulation and Data Analysis Workshop October 28–30, 2002.* Web. 9 June 2012. http:// cgwp.gravity.psu.edu/events/SrcSimDA/slides/Mezzacappa.pdf.

———. Personal Interview. 20 August 2003. Audiocassette.

Michaels, David and Celeste Monforton. "Manufacturing Uncertainty: Contested Science and the Protection of the Public's Health and Environment." *American Journal of Public Health* 95.S1 (2005): S39–S48. Print.

Morgan, Mary and Margaret Morrison. *Models as Metaphors: Perspectives on Natural and Social Science.* London: Cambridge University Press, 1999. Print.

Morris, John C. "Whither Fema? Hurricane Katrina and FEMA's Response to the Gulf Coast." *Public Works Management and Policy* 10.4 (2006): 284–294. Print.

Moser, Susanne C. "Communicating Climate Change: History, Challenges, Process and Future Directions." *WIREs Climate Change* 1 (2010): 31–53. Print.

Murray, Janet. *Hamlet on the Holodeck: The Future of Narrative in Cyberspace.* New York: Free Press, 1997. Print.

Myers, David. "The Aesthetics of the Anti-Aesthetics." *Online Proceedings from the Aesthetics of Play Conference.* 2005. Web. June 10, 2013. http://www.aestheticsofplay.org/myers.php.

Myers, Greg. "The Social Construction of Two Biologists' Proposals." *Written Communication* 2 (1985): 219–45. Print.

———. *Writing Biology: Texts in the Social Construction of Scientific Knowledge.* Madison: University of Wisconsin Press, 1990. Print.

National Energy Research Scientific Computation Center (NERSC). "2001 Annual Report." Berkeley: Berkeley Lab Technical and Electronic Information Department (2001). PDF.

Newman, Sara. "Aristotle's Notion of 'Bringing-Before-the-Eyes': Its Contributions to Aristotelian and Contemporary Conceptualizations of Metaphor, Style, and Audience." *Rhetorica* 20 (2002): 1–23. Print.

Norton, John D. "Are Though Experiments Just What You Thought?" *Canadian Journal of Philosophy* 26.3 (1996): 333–366. Print.

Oak Ridge National Lab (ORNL). March 7, 2006. Web. 9 May 2013. http://www.ornl.gov/.

Oak Ridge National Lab (ORNL). *Publications for Search Engines. March 6, 2006.* Web. 9 May 2012. http://ornl.gov/ornlhome/publications_listing_cppr.shtml.

Olin, Doris. *Paradox.* Quebec: McGill-Queens University Press, 2003. Print.

Owens, D., K. N. Lohr, D. Atkins, J. R. Treadwell, J. T. Reston, E. B. Bass, S. Chang, and M. Helfand. "AHRQ Series Paper 5: Grading the Strength of a Body of Evidence when Comparing Medical Interventions: AHRQ and the Effective Health Care Program." *Journal of Clinical Epidemiology* 63.5 (2010): 513–523. Print.

Panoff, Robert M. and Michael J. South. "The Role of Paradox in Science and Mathematics." *SSEP eLabtric Notebook.* 23 October 2003. Web. 6 March 2006. http://www.shodor.org/ ssep/pae/editorials/paradox.html.

Paul, Danette. "In Citing Chaos: A Study of the Rhetorical Use of Citations." *Journal of Business and Technical Communication* 14.2 (2000): 185–222. Print.

Paul, Danette, Davida Charney, and Aimee Kendall. "Moving Beyond the Moment: Reception Studies in the Rhetoric of Science." *Journal of Business and Technical Communication,* 15.3 (2001): 372–399. Print.

Pearce Fred."How the 'Climategate' Scandal is Bogus and Based on Climate Sceptics' Lies." *The Guardian* 9 Feb. 2010: n. pag. Web. 13 March 2013. http://www.theguardian.com/environment/2010/feb/09/climategate-bogus-sceptics-lies.

Peirce, Charles Saunders. *Collected Papers of Charles S. Peirce*. Charlottesville: InteLex Corporation, 2013. Web. 8 March 2013. http://www.nlx.com/collections/95.

———. *The Essential Peirce: Selected Philosophical Writings (1867–1893)*. Eds. Nathan Houser and Christian Kloesel. Bloomington: Indiana University Press, 1992. Print.

Peterson, Arthur C. *Simulating Nature: A Philosophical Study of Computer-Simulation Uncertainties and Their Role in Climate Science and Policy Advice*. Netherlands: Chapman and Hall/CRC, 2012. Print.

Piltz, Rick. "Open Letter to the U.S. Government from U.S. Scientists on Climate Change and the IPCC Reports." *Climate Science Watch*. Climate Science Watch, 2010. Web. 10 June 2013. http://www.climatesciencewatch.org/2010/03/11/open-letter-to-the-u-s-government-from-u-s-scientists-on-climate-change-and-the-ipcc-reports.

Popper, Karl. *The Logic of Scientific Discovery*. Trans. Logik der Forschung. London: Hutchinson, 1959. Print.

Quine, Willard Van Orman. *The Ways of Paradox and Other Essays, Revised Edition*. Cambridge: Harvard University Press, 1976. Print.

Randerson, James. "Climate Researchers 'Secrecy' Criticised—But MPs Say Science Remains Intact." *The Guardian* March 30, 2010: n. pag. Web. 13 March 2013. http://www.theguardian.com/environment/2010/mar/31/climate-mails-inquiry-jones-cleared.

Ricoeur, Paul. *The Rule of Metaphor: The Creation of Meaning in Language*. Oxford: Psychology Press, 2003. Print.

Romm, Joe. "Let's look at one of the illegally hacked emails in more detail—the one by NCAR's Kevin Trenberth on "where the heck is global warming?" Climate Progress. 21 November 2009. Web. 3 June 2013. http://thinkprogress.org/climate/2009/11/21/204991/hacked-emails-ncar-kevin-trenberth/.

Rorty Richard. *Essays on Heidegger and Others: Philosophical Papers. Vol. 2*. Cambridge: Cambridge University Press, 1991. Print.

———. *Consequences of Pragmatism*. Minneapolis: University of Minnesota Press, 1982. Print.

Rowland, F. "Methods and Motives for Publishing Original Work in Science." *Communicating Science: Professional Contexts*. Eds. Eileen Scanlon, Roger Hill and Kirk Junker. New York: Routledge, 1999: 61–71. Print .

Rundblad, Gabriella. "Impersonal, General, and Social: The Use of Metonymy Versus Passive Voice in Medical Discourse." *Written Communication* 24.3 (2007): 250–277. Print.

Russell, Bertrand. *The Principles of Mathematics*. Cambridge: Cambridge University Press, 1903. Print.

Rymer, Jone. "Scientific Composing Processes: How Eminent Scientists Write Journal Articles." *Advances in Writing Research,Vol. 2*. Ed. David Jolliffe. Norwood, NJ: Ablex, 1988. 211–250. Print.

Saad, Lydia. "Americans' Concerns about Global Warming on the Rise." *Gallup Politics*. Gallup, 8 April 2013. Web. 8 May 2013. http://www.gallup.com/poll/161645/americans-concerns-global-warming-rise.asp.

Saussure, Ferdinand de. *Course in General Linguistics*. Eds. Charles Bally and Albert Sechehaye. Trans. Wade Baskin. London: Owen, 1960. Print.

Schneider G. M., D. Schwalbe, and T. M. Halverson. "Teaching Computational Science in a Liberal Arts Environment." *SIGCSE Bulletin* 30.2 (1998): 57–60. Print.

Schmidlin, Thomas W. "On Evacuation and Deaths from Hurricane Katrina." *Bulletin of the American Meteorological Society* 87.6 (2006): 754–756. Print.

Scientific Discovery through Advanced Computing (SciDAC). 02 Jan. 2013. Department of Energy. 8 May 2012. Web. 4 April 2013. http://www.scidac.gov.

Segelken, Roger. *Bumblebees Finally Cleared for Takeoff: Insect Flight Obeys Aerodynamic Rules, Cornell Physicist Proves*. Ithaca: Cornell University News, 20 March 2000. Web. 3 April 2013. http://web.archive.org/web/20000511054430/http:/www.news.cornell.edu/releases/March00/APS_Wang.hrs.html.

Shapin, Steven. "Pump and Circumstance: Robert Boyle's Literary Technology." *Social Studies of Science* 14. 4 (1984): 481–520. Print.

Sherman, William R. and Alan B. Craig. *Understanding Virtual Reality: Interface, Application, and Design.* San Francisco: Morgan Kaufmann, 2003. Print.

Shiflet, Angela B. "Computer Science with the Sciences: An Emphasis in Computational Science." *SIGCSE Bulletin* 34.4 (2002): 40–43. Print.

Sokal, Alan. "What the Social Text Affair Does and Does Not Prove." *Critical Quarterly* 40.2 (1998): 3–18. Print.

"Solving the Mystery of the Missing Neutrinos." *Nobelprize.org.* Nobel Media AB 2013. Web. 9 May 2013. http://www.nobelprize.org/nobel_prizes/themes/physics/bahcall/.

Soon, William, Sallie Baliunas, Arthur Robinson, and Zachary Robinson. "Environmental Effects of Increased Atmospheric Carbon Dioxide." *Journal of American Physicians and Surgeons* 12 (2007): 79–90. Print.

Sorenson, Roy A. *Thought Experiments.* New York: Oxford University Press, 1992. Print.

Spinellis, Diomidis. "Reading, Writing and Code." *ACM Queue* 1.7 (2003): 84–89. Print.

Spinuzzi, Clay. "Toward Integrating Our Research Scope: A Sociocultural Field Methodology." *Journal of Business and Technical Communication* 16.1 (2002): 3–32. Print.

———. "Pseudotransactionality, Activity Theory, and Professional Writing Instruction." *Technical Communication Quarterly* 5.3 (1996): 295–308. Print.

———. "Towards a Hermeneutic Understanding of Programming Languages." *Currents in Electronic Literacy* 6 (Spring 2002): n. pag. Web. 4 May 2013. http://currents.cwrl.utexas.edu/spring02/spinuzzi.html.

———. *Tracing Genres through Organizations: A Sociocultural Approach to Information Design.* Cambridge, MA: MIT Press, 2004. Print.

Spoel, Philippa et al. "Public Communication of Climate Change Science: Engaging Citizens through Apocalyptic Narrative Explanation." *Technical Communication Quarterly* 18.1 (2009): 49–81. Print.

Swales, John M. *Genre Analysis.* Cambridge: Cambridge University Press, 1990. Print.

Syverson, Margaret. *The Wealth of Reality: An Ecology of Composition.* Carbondale: Southern University Press, 1999. Print.

Tannen, Deborah. *The Argument Culture: Stopping America's War of Words.* New York: Ballantine Books. 1999. Print.

Tarone, Elaine, Sharon Dwyer, Susan Gillette, and Vincent Icke. "On the Use of the Passive and Active Voice in Astrophysics Journal Papers: With Extensions to Other Languages and Other Fields." *English for Specific Purposes* 17:1 (1998): 113–132. PDF file.

United States House of Representatives. *A Failure of Initiative: Final Report of the Select Bipartisan Committee to Investigate the Preparation for and Response to Hurricane Katrina.* 15 February 2006. Washington DC: U.S. Government Printing Office. PDF file.

United States Senate. Committee of Homeland Security and Governmental Affairs. *Preparing for a Catastrophe: The Hurricane Pam Exercise. Hearing before the Committee on Homeland Security and Governmental Affairs.* 24 January 2006. Washington DC: U.S. Government Printing Office. PDF file.

van Heerden, Ivor and Mike Bryan. *The Storm: What Went Wrong and Why During Hurricane Katrina: The Inside Story from One Louisiana Scientist.* Penguin: New York, 2006. Print.

van Veuren, P. J. J. "Against Rhetoricism." *South African Journal of Philosophy* 16.4 (1997): 158–166. PDF file.

Vande Kopple, William J. "Some Characteristics and Functions of Grammatical Subjects in Scientific Discourse." *Written Communication* 11.4 (1994): 534–564. PDF file

Varadarajan, Navin. Personal Interview. March 30, 2004.

Varadarajan, Navin, Jongsik Gam, Mark J. Olsen, George Georgiou, and Brent L. Iverson. "Engineering of Protease Variants Exhibiting High Catalytic Activity and Exquisite Substrate Selectivity." *Proceedings of the National Academy Of Science of the United States of America* 102.19 (2005): 6855-6860. Web. April 10, 2013. http://www.pnas.org/content/102/19/6855.abstract.

Vatz, Richard. "The Myth of the Rhetorical Situation." *Philosophy and Rhetoric* 6 (1973): 154–161. Print.

Vygotsky, Lev S. "Imagination and Creativity in Childhood." *Journal of Russian and East European Psychology* 42.1 (2003) 7–97. Print.

Walczak, Monica. "The Classical Conception of Rationality." *The Paideia Project On-line.* Boston University. 10 June 2000. Web. 7 March 2012. http://www.bu.edu/wcp/Papers/TKno/TKnoWalc.htm.

Walli, Ron. *ORNL Heads DOE Project that Looks to the Stars.* Oak Ridge: Oak Ridge National Laboratory. 2001. Web. 5 June 2013 http://web.ornl.gov/info/press_releases/get_press_release.cfm?ReleaseNumber=mr20010918-00.

Walsh, Lynda. *Scientists as Prophets: A Rhetorical Genealogy.* New York: Oxford University Press. 2013. Print.

Walsh, Lynda. "Before Climategate: Visual Strategies to Integrate Ethos Across the 'Is/Ought' Divide in the IPCC's Climate Change 2007: Summary for Policy Makers." *Poroi* 6.2 (2009): Article 4. Web. 9 June 2013. http://ir.uiowa.edu/cgi/viewcontent.cgi?article=1066&context=poroi.

Weissman, Laurence M. "A Methodology for Studying the Psychological Complexity of Computer Programs." Diss. University of Toronto, 1974. Ann Arbor: UMI, 1974. PDF file.

Williams, Bernard. Interview by Andrew Marr. *Start of the Week.* BBC Four. London. 18 November 2002. Radio.

Williams, Stewart. "Rethinking the Nature of Disaster: From Failed Instruments of Learning to a Post-Social Understanding." *Social Forces* 87.2 (2008): 1115–1138.

Wilson, Kris M. "Communicating Climate Change through the Media: Predictions, Politics and Perceptions of Risk." *Environmental Risks and the Media.* Eds. S. Allan, B. Adam and C. Carter. New York: Routledge, 2000. 201–217. Print.

Winsberg, Eric. *Science in the Age of Computer Simulation.* Chicago: University of Chicago Press, 2010. Print.

———. "Simulations, Models, and Theories: Complex Physical Systems and Their Representations." *Philosophy of Science* 8 (2001): S442–54. Print.

———. "Simulated Experiments: Methodology for a Virtual World." *Philosophy of Science* 70 (2003): 105–125. Print

———. "Simulation and the Philosophy of Science: Computationally Intensive Studies of Complex Physical Systems." Diss. Indiana University, 1999. Ann Arbor: UMI, 1999. PDF file.

———. "Computer Simulation and the Philosophy of Science." *Philosophy Compass* 4.5 (2009): 835–845. Print.

Witte, Stephen P. "Context, Text, Intertext: Toward a Constructivist Semiotic of Writing." *Written Communication* 9.2 (1992): 237–308. Print.

Wolf, Mark. "Subjunctive Documentary: Computer Imaging and Simulation." *Collecting Visible Evidence.* Eds. Jane Gaines and Michael Renov. Minneapolis: University of Minnesota Press, 1999. Print.

Wolfe, Joanna. "Why the Rhetoric of CS Programming Assignments Matters." *Computer Science Education* 14.2 (2004): 147–163. Print.

Woolfson, Michael M. and G.J. Pert. *An Introduction to Computer Simulation.* New York: Oxford University Press, 1999. Print.

Wordsworth, William and Samuel Taylor Coleridge. *Wordsworth's Preface to Lyrical Ballads.* British Library, 1987. Print.

Index

CPSIA information can be obtained at www.ICGtesting.com
Printed in the USA
BVOW07*1632011213

337569BV00003B/6/P

9 780739 175569